D0812648

HUMAN FACTORS IN QUALITY ASSURANCE

WILEY SERIES IN HUMAN FACTORS

Edited by David Meister

Human Factors in Quality Assurance

Douglas H. Harris and Frederick B. Chaney

Human Factors in Quality Assurance

DOUGLAS H. HARRIS
FREDERICK B. CHANEY
North American Rockwell Corporation

JOHN WILEY & SONS, INC.
New York • London • Sydney • Toronto

10 9 8 7 6 5 4 3 2 1

Library of Congress Catalog Card Number: 68-57099
SBN 471 35313 2
Printed in the United States of America

SERIES PREFACE

Technology is effective to the extent that men can operate and maintain the machines they design. Equipment design which consciously takes advantage of human capabilities and constrains itself within human limitations amplifies and increases system output. If it does not, system performance is reduced and the purpose for which the equipment was designed is endangered. This consideration is even more significant today than in the past because the highly complex systems that we develop are pushing human functions more and more to their limits of efficient performance.

How can one ensure that machines and machine operations are actually designed for human use? Behavioral data, principles, and recommendations—in short, the Human Factors discipline—must be translated into meaningful design practices. Concepts like ease of operation or error-free performance must be interpretable in hardware and system terms.

Human Factors is one of the newer engineering disciplines. Perhaps because of this, engineering and human-factors specialists lack a common orientation with which their respective disciplines can communicate. The goal of the Wiley Human Factors Series is to help in the communication process by describing what behavioral principles mean for system design and by suggesting the behavioral research that must be performed to solve design problems. The premise on which the series is based and on which each book is written is that Human Factors has utility only to the degree that it supports engineering development; hence the Series emphasizes the *practical application* to design of human-factors concepts.

Because of the many talents on which Human Factors depends for its implementation (design and systems engineering, industrial and experimental psychology, anthropology, physiology, and operations research, to name only a few), the Series is directed to as wide an audience as possible. Each book is intended to illustrate the usefulness of Human Factors to some significant aspect of system development, such as human factors in design or testing or simulation. Although cookbook answers are not provided, it is hoped that this pragmatic approach will enable the many specialists concerned with problems of equipment design to solve these problems more efficiently.

David Meister
Series Editor

PREFACE

This book is addressed to meeting quality objectives through the performance of people in industry. It is based on the assumption that the decisions and actions of people are the most important considerations in the attainment of quality objectives. The importance of people is evident both in the production process in which errors lead to defective products and in the quality assurance process in which errors lead to inaccurate information and ineffective problem solving.

In this book people are viewed as critical elements in sociotechnical systems. The size of the system may be as large as a multidivisional corporation or as small as a group of assembly workers with their tools and procedures. The people, equipment, and information organized to ensure that products are made to certain standards of quality may be thought of as a quality system. Within this framework the improvement of human performance is achieved through changes in people, equipment or information. We refer to this approach, which considers all elements in the system, as human factors research and engineering. Information obtained by scientific methods is put to use by changing the elements of the system or by modifying the way in which the elements interact with one another. These changes lead to improved human performance and ultimately to increased system effectiveness.

The book was written to serve two purposes. The primary purpose is to provide a source of information and ideas for supervisors, engineers, and specialists who are concerned with quality objectives in industrial operations. The secondary purpose is to provide material useful in the education of professional people such as industrial psychologists, industrial engineers, and human factors specialists. To this end the content and organization of the book have been developed to provide the reader with an understanding of human behavior in industry as well as with specific ideas and techniques that may be used to improve industrial operations. The book is organized to provide, first of all, a frame of reference for relating human factors research and development to quality assurance objectives. Then the jobs of the quality supervisor, engineer, and inspector are discussed in light of the factors that affect their performance. Finally, approaches and techniques that may be used to improve human performance in quality assurance operations are described.

Human Factors research and development efforts conducted over the last five years at the North American Rockwell Corporation form the base of this book. Although we have drawn from the experience of other investigators in order to provide a comprehensive coverage of the field, our point of view has been shaped largely by our own efforts in the solution of quality problems. Much of the credit for the success and continuity of these efforts belongs to James L. Thresh, Manager of Navigation Systems Quality Assurance, Autonetics Division of North American Rockwell Corporation. It was his investment of funds, time, and suggestions that enabled these efforts to get started and grow to the extent that their contributions could be evaluated. We also recognize that the efforts of many people made this book possible. Human factors research and development studies usually place additional burdens on supervisors and the people in their organizations. However, when the short-term burden was viewed as a trade-off for a potential long-term company gain, the added work was willingly accepted. Consequently, nearly all the studies credited to ourselves in this book were actually joint efforts in which we played only a part. We wish also to recognize the technical direction and support provided by Kenneth S. Teel, Autonetics' Human Factors Manager, in these research efforts and in the preparation of the book.

Douglas H. Harris
Frederick B. Chaney

Anaheim, California
November, 1968

CONTENTS

HUMAN FACTORS IN QUALITY ASSURANCE

I

HUMAN PERFORMANCE IN INDUSTRY

In the past, when most products were handmade, human performance was obviously very important, but as machines began to take over more jobs the role of people in industry became somewhat obscured. Today, with the rapid advance of automation, human performance appears to be losing much of its importance. As a result of this trend the modern manager must seriously consider some basic questions. To what extent do people still influence corporate profits and new business? How much capital should be invested in improving human performance as opposed to investments in automation such as computerized data systems?

While these are obviously difficult questions, there is plenty of evidence to indicate that people still play a major role in determining product quality and reliability. An analysis of 23,000 defects found in testing components for nuclear weapon systems indicated that 82 percent, or 19,200 of these defects, were directly attributable to human error [1]. A survey of several major missile systems indicated that 20 to 50 percent of the reported equipment failures were due to human errors [2]. These examples dramatically illustrate the continuing importance of human performance in industry.

Many other examples could be cited that also illustrate the critical role of people in quality assurance. People conceive and design new products and processes. People select the materials, parts, and equipment required for new products. People assemble, inspect, and test. Finally, people package, deliver, and service these products. In performing all these functions, as in most human activities, each task is subject to various types of human error. This book provides a summary of information about the behavior of people that may be used to minimize the chances of human error. The systematic use of this information should lead to reduced costs and products that consistently meet quality standards. The specific objective of this first chapter is to discuss human performance in terms of the various activities which typically are found in industry and to consider some of the major factors which influence this performance. The current state-of-the-art in the classification and prediction of human error is reviewed briefly. Particular emphasis is placed on the detection and correction of workmanship errors in the production system.

TYPES OF HUMAN BEHAVIOR

One of the first requirements for a systematic study of human performance and error in industry is a standard set of terms for classifying this behavior. The need for such a task taxonomy has been clearly stated by Miller [3] and additional contributions in this area have been made by Meister and Rabideau [4]. A classification of human performance which is particularly suited to the industrial situation was provided by Altman [5]. A slightly modified version of this classification scheme is presented in Table 1.1. This system groups very broad functions of human performance under such phases as planning, designing and developing, producing, distributing, operating, and maintaining. A much more detailed classification system that could be used to describe the behavior for a specific function has been provided by Berliner [6]. As shown in

Table 1.1 Classification of Human Performance in Industry

- Planning
    - Definition of objectives
    - Identification of capabilities
    - Definition of constraints
    - Evaluation of tradeoffs
    - Establishment of requirements
    - Scheduling
- Designing and Developing
    - Analysis and simulation
    - General, functional, or conceptual design
    - Detailing and checking
    - Mockup and prototype fabrication
    - Developmental testing
- Producing
    - Fabricating and assembly
    - Handling and transporting
    - Inspecting and checking
- Distributing
    - Selling and packaging
    - Transporting and installing
    - Installation inspection and checking

Table 1.2, the major categories for this system are perceptual, mediation, communication, and motor processes.

In addition to an adequate classification system, the study of human performance in industry also requires procedures for quantifying each of the specific types of behavior. This requirement can generally be satisfied by some combination of time and error scores. In this book techniques for determining the frequency and importance of errors of commission and omission are particularly important. The collection and use of this type of information is one of the primary functions of quality assurance. In developing a quality assurance program to minimize the probability of human error and to detect effectively and eliminate the errors that may

Table 1.2 Classification of Behaviors

Processes	Activities	Specific Behaviors
Perceptual processes	Searching for and receiving information	Observes Reads Receives
	Identifying objects, actions, events	Discriminates Identifies Locates
Mediational processes	Information processing	Codes Interpolates Translates
	Problem solving and decision making	Calculates Compares Estimates
Communication processes		Answers Directs Instructs Requests Transmits
Motor processes	Simple/discrete	Closes Connects Moves Sets
	Complex/continuous	Adjusts Aligns Regulates Tracks

occur in the production of a new system, it is important to consider the various factors which may influence human performance.

FACTORS IN HUMAN PERFORMANCE

The quality and output of human performance is determined by individual characteristics, the physical environment, and the task information provided to the employee. In general, the term used for these influences is human factors. These factors individually and in combination determine how well people will do their jobs. One of the major misconceptions that appears to be prevalent in industry today is the belief that most human performance problems can be cured through improved training and/or motivation programs. Much of this current emphasis on improving the quality of a product through people-oriented programs is based on the false belief that human error results primarily from bad attitudes and lack of attention. The assumption seems to be that if people would just try harder, performance could be free of error. In general, research has indicated that the changes produced through improved attitudes and motivation tend to be limited and transitory. Even programs aimed at more basic improvements in human performance through better selection and training techniques often fail to have the desired level of impact. Substantial and permanent improvements in human performance can be obtained, however, through applying human factors principles to the development and evaluation of new tools, techniques, equipment, and job aids. The potential for improvement through changes in the job and its environment appears to be much greater than the results obtained from motivational efforts [7].

A view of human performance which provides a scientific basis for improvements in quality and output is that errors are a result of an interaction between people and other elements in the production system such as equipment, workspace layout, hand tools, and job instructions. This approach appears to be more likely to provide sustained improvement because it places the responsibility for performance improvement on the shoulders of both employees and company management. In terms of quality improvement this system concept of error accepts the reality of human fallibility and attempts to identify the specific changes needed to reduce the probability of human error in critical operations. This would certainly appear to be a more scientific approach than striving for perfect human performance by encouraging people to be alert and try harder. A systems view of human error also avoids such meaningless questions as "What is good workmanship?" and directs attention to the practical problem of identifying and reducing the causes of error.

CLASSIFICATION AND DIAGNOSIS OF ERROR

One of the primary problems with current quality assurance technology is that the defect description and classification systems describe production errors in terms such as wrong part, poor solder, broken lead, and other combinations of nouns and adjectives which are descriptive but contain very little information on defect causes. By classifying these defects in terms of the information, environment, and behavior which produce them, rather than simply describing the defect, it is possible to obtain the type of diagnostic information needed for quality improvement.

A number of alternative error classification approaches may be summarized under five basic headings: performance information, physical situation, individual considerations, error impact, and corrective action. The relationship among these five areas and the basis for classifications are summarized in Figure 1.1.

Performance information classifications tend to emphasize the activity in which the workers engage at the time an error occurs. Physical situation approaches deal with the physical and operational environments that increase the probability of error. Classifications dealing with the individual considerations tend to emphasize the motivation and ability of the operator at the time an error was committed. Error impact classification systems deal with the chances of finding the error, ease of correction, and the seriousness of the error if it is undetected. This system of error classi-

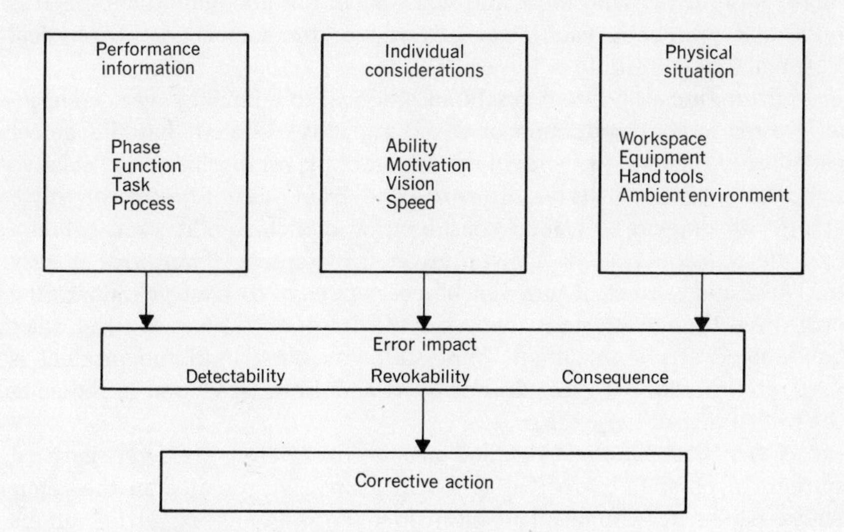

Figure 1.1 Basis for classification and diagnosis of error.

fication is based on the approach outlined by Altman [8]. All of these approaches to error classification provide information which has some diagnostic value and may be useful in determining the most appropriate form of corrective action. The potential uses for each type of classification system are summarized.

1. *Performance Information.* The first step in the diagnosis of any error should be an accurate determination of the type of performance specified and a description of the behavior that actually occurred. Without this information it is impossible to determine whether the error was due to improper information or a failure to perform a required act.

2. *Individual Considerations.* When errors occur, even though adequate performance information is provided, we must consider the characteristics of the individual performing the task. Information should be obtained on the task requirements in terms of ability, motivation, vision, speed, and other individual characteristics. In some instances it will be possible to obtain the necessary changes in these characteristics by adequate employee training. Longer range solutions may involve the selection and placement of individuals with better qualifications for a particular task or redesigning the job to eliminate human tasks that may be more adequately performed by machines.

3. *Physical Situation.* Broadly conceived, the physical situation includes all the factors except information and people that may contribute to error. The specific categories include such factors as the ambient environment, equipment, hand tools, and workspace. For adequate diagnosis it is important to classify each error in terms of the factors in the physical situation that contribute to this error.

4. *Error Impact.* Since it is seldom practical to eliminate every conceivable error in the production of a system, it is essential that the defect classification technique provide a method of establishing the relative importance of various types of errors. The appropriate form of corrective action will depend on a joint consideration of detectability, revocability, and the consequences of a given error; for example, if an error is easy to detect and correct, it may not be cost effective to try to eliminate the error even though its consequences could be quite serious. On the other hand, some errors are almost impossible to detect until the product is used, and therefore a great deal of effort should be expended in reducing the probability of these errors.

5. *Corrective Action.* A detailed scheme for relating error classification systems to possible corrective action measures is presented in the referenced paper by Altman. The basic idea is that once an error can be classified in terms of impact, performance information, individual char-

acteristics, and the physical situation it is possible to identify one or more actions that will reduce the chances of this type of error in the future.

It is important to realize that the "cause of the error" and the most economical means for avoiding the error in the future may be quite different; for example, a high error rate in electronic wiring operations may be attributed to such causes as training or low motivation. However, the amount of training required to obtain the necessary skill level and the increased incentives needed to raise the level of motivation may be very expensive. Under these conditions the most economical methods for reducing the error rate may be improvements in work instructions, better tools and equipment, color coding, or detailed visual aids for critical areas. On the other hand, defects that occur during complex chemical processes may be attributed by production people to poorly defined process instructions and improper equipment. After adequate investigation, it may be that many of these errors can be eliminated by brief on-the-job training for certain critical steps in the process and by providing error feedback to each of the operators. These examples attempt to illustrate the fallacy of corrective action efforts that are aimed at identifying "the cause" of any major defect. The alternative approach is to identify the specific behavior and other factors that are associated with critical types of error and then determine the combination of treatments that will economically reduce the chances of these errors in the future. In addition to its demonstrated value in corrective action situations, this approach may also be used in minimizing the probability of human error in new situations. Under these conditions the limiting factor is the feasibility of predicting human error rates for specific jobs and job situations.

PREDICTION OF TASK RELIABILITY

The data store developed at the American Institute for Research by Payne and Altman [9] is the first and perhaps best known attempt to predict human reliability. In general, this data store provides success probabilities for a large number of typical task elements. The basic assumption underlying the system is that the reliability of complex tasks can be estimated by combining the probabilities of success for each of the elements in that task. To use the data store the success probabilities for the task elements are multiplied together to obain a single prediction of task reliability. This approach can be illustrated by the following example. Consider a production operator who has a simple job of switching a filter from a low pressure to a high pressure system. The task reliability and the success probabilities for the four discrete elements of this task are shown in Table 1.3.

Table 1.3 Sample Task Reliability Determination

Task Reliability	Success Probability	Task and Elements
0.9797		Switch the filter from the low-pressure to the high-pressure system:
	0.9984	(a) Remove one fitting with two wrenches.
	0.9952	(b) Fasten filter to bulkhead fitting.
	0.9930	(c) Fasten new filter to high-pressure fitting.
	0.9930	(d) Fasten filter from pneumatics line to vehicle.

Swain [10] has shown that the successive multiplication of probabilities leads to misleading results in many complex systems. Under these conditions he has suggested the use of a procedure for estimating conditional probabilities of larger unit behavior and then using these units in a multiplicative fashion to estimate task reliabilities.

Although the qualification and prediction of human reliability are worthwhile objectives, the required technology is still in the developmental stages. Some of the current obstacles in obtaining a practical system for determining task reliability include (a) the complexity and subjectivity of present techniques, (b) the difficulty in meeting assumptions required to justify the use of these techniques, and (c) the need for an improved data bank for human performance [11]. In spite of the problems currently associated with trying to predict human reliability in production processes, there are a number of things which can be done to improve the identification and reduction of human error in industrial performance.

WORKMANSHIP ERRORS

As defined by Meister [12], a workmanship error is a discrepancy between a standard and the hardware produced by a worker during the production process. The standard might consist of a blueprint, a visual aid, or an operator instruction document and the item produced might be anything from a metal casting to a complex integrated circuit the size of a pinhead. Defects that are typically classified as workmanship errors include such things as wrong part, reverse polarity, stripped thread, mislocated hole, insulation burned, loose connection, and damaged component. Since one of the main objectives of quality assurance is the identification and reduction of workmanship errors, it is reasonable to question the importance of these types of errors. The opening paragraph of this chapter cited two examples which illustrated the proportion of total defects which are typically attributed to human performance. An analysis reported by Meister [13] for failures on an Atlas missile testing site showed that 22 percent of them were determined to be human initiated. It is important to recognize that this 22 percent was identified after a large number of human-initiated failures had been screened out by prior inspection and acceptance testing of equipment. In general, about 80 percent of the defects found during the production of complex systems is attributed to human error. In terms of their effect on operational reliability, an evaluation of Atlas missile system performance indicated that human-initiated failures contributed 24 percent of the total unreliability of the system. These examples clearly illustrate the importance of workmanship error in terms of cost, safety, and system performance factors. When we consider the fact that one workmanship error in a modern missile system can result in the loss of numerous lives and millions of dollars, it seems safe to conclude that this type of human performance is not only important, but is perhaps one of the most critical variables in modern industry. Recognition of the importance of workmanship errors has focused increased attention on the need for better defect information systems.

DEFECT DETECTION AND REPORTING

While quality assurance has grown to incorporate a number of new disciplines, product inspection is still the most critical activity performed by this function. The inspection process is the basic source of information on workmanship and other types of error. As a result, the effectiveness of the total quality system is highly dependent on the accuracy of the basic quality data that are collected and reported by the inspector. If the inspection process were 100 percent accurate, workmanship errors probably would not represent a serious problem in the field use of products and their impact would be felt primarily in terms of rework and scrap costs within the manufacturing organization. The information to date, however, clearly indicates that inspection is not 100 percent accurate.

Discussions and interviews with experts who have worked in the field of product inspection for a number of years generally indicate that the inspector is, on the average, at least 90 percent accurate. Data reported by McCornack [14] indicate that average inspection effectiveness is about 85 percent, but more recent research has shown that overall statements about inspection accuracy are relatively meaningless, since variations due to product complexity alone can cause a range of average inspection accuracies of 20 to 80 percent. When we consider other variables such as the actual defect rate, the inspection environment, and the types of inspection methods employed, a much wider range of variability is found. Based on detailed job sample measures of inspection accuracy in a variety of functions, it is seldom that we find over 50 to 60 percent of the defects being detected at any point in time by a single inspector. Fortunately, there are a number of specific techniques that can be used to improve inspection performance and the effectiveness of these techniques can be assessed with a high degree of precision. These techniques for improving inspection accuracy and appropriate methods for assessing their impact constitute a large portion of the material contained in this book.

A second major problem associated with the inspection function is that of summarizing and reporting information on workmanship error in an economical fashion. To be useful workmanship data must be summarized in terms of meaningful error rates and reported to those individuals who are responsible for reducing these error rates. The psychologist is frequently asked to determine what level of workmanship error is to be expected for various operations to provide a base for evaluating current performance. Theoretically this type of information could be developed using the data store approach just discussed. However, accurate information on the population of production behavior elements is not available and probably will not be developed for some time. Even if this type of standard were available, it would be of limited value due to some of the practical problems involved in computing meaningful workmanship error rates in the operational environment.

One critical problem in measuring workmanship is the development of a meaningful base. A meaningful error rate for any given operation could be computed by dividing the number of observed errors by the total number of possible errors. It is seldom practical, however, to calculate the total number of possible errors and all methods currently used to estimate this base are subject to various sources of bias. Typical approaches that are used in reporting quality statistics include (a) defects per thousand hours, (b) defects per inspection operations, (c) defects per part, (d) percent defective, and (e) defects divided by some complexity factor. Although these types of index may well provide useful management information, they do not supply the data required to make meaningful statements about human error rates in the production environment. As a result, inspection information should be viewed as a means of diagnosing problem areas and not of making statements about the over-all adequacy of human performance.

Once this basic point is understood and accepted, a quality information system can be developed that will serve as a useful tool for monitoring the production process and identifying general workmanship problem areas. After a quality information system has been developed that provides rapid and accurate identification of general problem areas, attention can then be shifted to the appropriate techniques for detailed diagnosis and correction of quality problems.

QUALITY IMPROVEMENT

Improving a modern production process is a very difficult and complex job requiring the use of many different types of information and a variety of technical specialties. Perhaps the apparent difficulty of this type of activity and the hard-nosed, pragmatic viewpoint of production people have discouraged human factors specialists from devoting a significant amount of effort to the study of human performance in production systems, but the critical role of people in the production process and the potential cost savings from improved human performance should serve as strong incentives for behavioral specialists in the future. Previous attempts to improve human performance have been severely limited by their emphasis on a particular technique and the segregation of psychological and technical approaches to improving the industrial process.

One of the most promising new developments in the study of industrial management is the increased emphasis on collaboratoin between behavioral science and technology [15]. This approach requires a joint consideration of the engineering and psychological factors that influence productivity. The concept is to establish *sociotechnical systems* that will enable employees and technology to produce the best results. To be effective this collaboration should start with the initial conception of a new product and continue through the design, manufacturing, testing, and evaluation phases. In established production systems collaboration between human factors experts and technical specialists is essential for obtaining the information required to identify potential areas for improvement. Psychological approaches including personnel selection, training, and motivation have produced significant improvements in the past. The most significant and permanent improvements, however, generally come from new methods and techniques for doing the job.

The primary contributions of human factors technology to the continuing evolution of process improvements are as follows:

1. Development of accurate information systems for identifying the weaknesses in current production systems.

2. Improvement of communications between production personnel and the technical specialists who develop new tools and techniques.

3. Systematic evaluation of alternative methods for job improvement to determine their impact on human performance before major expenditures are made.

4. Assistance in the application of participative techniques so that technical improvements will have the type of employee acceptance required to realize their potential benefits.

5. Development of a practical participative management technique required to sustain high quality performance and assure the continuing development of technical improvements.

This book attempts to show how human factors research can be used to translate the concept of a *sociotechnical system* into practical techniques for quality improvement. Most of the studies reported in the following chapters were conducted by individuals with training in both physical sciences and psychology in conjunction with experts in quality assurance technology. It is hoped that this book will clearly illustrate the potential gains from this joint effort and lead to increasing collaboration between human factors and quality assurance personnel.

SUMMARY

1. In spite of continuous improvements in technology and the rapid advance of automation, human performance is still a critical factor in determining product quality and reliability. The continuing importance of human performance is illustrated by the fact that 82 percent of the defects found in a study of nuclear weapon systems was directly attributable to human error.

2. Through the systematic study of human performance in industry, it has been possible to identify a number of specific factors that contribute to human error. Use of these findings should lead to reduced costs and products that more consistently meet quality standards.

3. One of the first requirements for the systematic study of human performance is a set of standard terms for describing behavior. A general task taxonomy for human performance in industry is provided in terms of phases, functions, processes, activities, and specific behaviors.

4. The effectiveness of human performance is determined by individual characteristics, the physical environment, and the task information provided to the employee. Potential improvements in performance through changes in the design of the job appear to be greater than potential improvements from training and motivational efforts.

5. Most defect classification systems do not provide the type of diagnostic information needed for quality improvement. The information needed for the classification and diagnosis of errors is provided with the following basic elements: (a) performance information, (b) physical situation, (c) individual consideration, (d) error impact, and (e) corrective action.

6. Significant attempts have been made to predict the reliability of specific human tasks by combining the success probabilities for a number of discrete behavior elements. Although the prediction of human reliability in production processes is a worthwhile goal, the required technology is still in the developmental stages.

7. A workmanship error is defined as a discrepancy between a standard and the hardware produced during the production process. About 80 percent of the defects found during the production of complex systems may be attributed to workmanship errors.

8. The significance of workmanship errors is increased by the fact that inspection accuracy is much lower than people generally assume it to be. Product complexity alone can cause inspection accuracy to vary from 20 to 80 percent. Job sample measures indicate that, on the average, a typical inspector seldom finds over 50 to 60 percent of the defects at any point in time. This book provides a summary of specific techniques which may be used to measure and improve the accuracy of inspection performance.

9. A meaningful workmanship error rate can only be computed by dividing the number of observed errors by the total number of possible errors. Since it is seldom practical to determine the total number of possible errors, inspection information should be viewed as a means of diagnosing problem areas and it should *not* be used to make statements about the overall adequacy of human performance.

10. One of the most promising developments in modern management is the increased emphasis on collaboration between behavioral science and technology. The study of human factors in quality assurance is an example of this new approach; it requires a joint consideration of both the engineering and psychological factors which affect productivity. The objective is to establish *sociotechnical systems* that will enable employees and technology to produce the best results.

REFERENCES

[1] Luther W. Rook, *Motivation and Human Error,* Report SC-TM-65-135. Sandia Corporation, Albuquerque, New Mexico, 1965.

[2] A. Shapero, J. I. Cooper, M. Rappaport, C. J. Erickson, K. H. Schaeffer and

C. J. Bates, Jr., Human Engineering Testing and Malfunction Data Collection in Weapon System Test Programs. Wright Air Development Division Technical Report 60–36, February 1960.

[3] R. B. Miller, Task Description and Analysis. In R. M. Gagne (Ed.), *Psychological Principles in System Development*. New York: Holt, Rinehart and Winston, 1962.

[4] D. Meister and G. F. Rabideau, *Human Factors Evaluation in System Development*. New York: Wiley, 1965.

[5] James W. Altman, Classification of Human Error. In W. B. Askren (Ed.), *Symposium on Reliability of Human Performance in Work*. AMRL-TR-67-68, Aerospace Medical Research Laboratories, Wright-Patterson Air Force Base, Ohio, May 1967.

[6] C. Berliner, D. Angell, and J. W. Shearer, *Behaviors, Measures and Instruments for Performance Evaluation in Simulated Environments*. Paper presented at the Symposium and Workshop on the Quantification of Human Performance. Albuquerque, New Mexico, August 17–19, 1964.

[7] Luther W. Rook, *Motivation and Human Error, loc. cit.*

[8] James W. Altman, Classification of Human Error, *loc. cit.*

[9] D. Payne and J. W. Altman, *An Index of Electronic Equipment Operability: Report of Development*. Pittsburgh, Pennsylvania: American Institute for Research Report AIR-C-43-1/62-FR, January 1962.

[10] Alan D. Swain, Some Limitations in Using the Simple Multiplicative Model in Behavior Quantification. In W. B. Askren (Ed.), *Symposium on Reliability of Human Performance in Work*. AMRL-TR-67-68, Aerospace Medical Research Laboratories, Wright-Patterson Air Force Base, Ohio, May 1967.

[11] Alan D. Swain, Human Factors in Design of Reliable Systems, Reprint SCR 748. Sandia Corporation, Albuquerque, New Mexico, February 1964.

[12] David Meister, Application of Human Reliability to the Production Process. In W. B. Askren (Ed.), *Symposium on Reliability of Human Performance in Work*. AMRL-TR-67-68, Aerospace Medical Research Laboratories, Wright-Patterson Air Force Base, Ohio, May 1967.

[13] David Meister, Analysis of Human Initiated Equipment Failures During Category I Testing, OSTF-1, Report REL R-054. General Dynamics/Astronautics, San Diego, California, 1961.

[14] R. L. McCornack, *Inspector Accuracy: A Study of the Literature*, Report SCTM 53-61 (14). Sandia Corporation, Albuquerque, New Mexico, 1961.

[15] R. C. Albrook, Participative Management: Time for a Second Look. *Fortune*, May 1967, 166–200.

II

THE NATURE OF QUALITY ASSURANCE

Quality in products and services has been an important quest in human activity from the time man began designing and producing things. However, the use of systematic methods to assure that products and services meet certain quality standards is a recent development. Twenty years ago quality assurance was the interest of a relatively small number of technical men in major companies. At the present time, however, there is a significant and increasing emphasis on quality. Quality assurance is the business of an increasingly large number of managers and professional people. There are several reasons for this change. Among them are the undertaking of national projects such as manned space missions in which widespread television and newspaper coverage make defects in quality visible to millions of people; the increased demand of consumers for quality products which has led to higher standards, such as 50,000 mile/five-year warranties on cars, and the realization by managers that money spent for research and engineering on quality problems can yield large profits in cost savings. The primitive, informal concepts of controlling quality therefore are being steadily replaced by systematic approaches to quality assurance.

The nature of quality and the objectives and approaches employed by quality assurance systems are discussed in this chapter. More complete presentations of quality assurance can be found in books by Juran [1] and Feigenbaum [2].

THE NATURE OF QUALITY

For purposes of quality assurance quality must be precisely defined. The term quality has no meaning in any absolute sense; it is meaningful only under certain specified conditions in the product cycle. These conditions generally involve some combination of the use to be made of the product and the product cost. Under the guidelines provided by product use and product cost quality standards can be established for specified product characteristics.

The Quality Characteristic

The basic element in quality is the quality characteristic. As discussed by Juran, a quality characteristic is a physical or chemical property, a dimension, a temperature, a pressure, or any other requirement used to define the nature of a product or service. Thus a metal cylinder may be defined approximately by such quality characteristics as type of metal, length, and diameter. A more precise definition would include as additional quality characteristics the hardness and finish of the cylinder surface. Similarly, a process for cleaning metals may be defined in terms of the chemical makeup of the solution, the temperature, and the time cycle for immersion.

The quality characteristic concept is useful in quality assurance because it enables the precise definition of product quality as the starting point for quality assurance. The quality of a product can be defined by the product's composite of quality characteristics. Acceptable quality, then, is a function of the extent to which the product's quality characteristics meet the standards established in light of customer satisfaction. Within this framework quality assurance becomes a matter of setting quality standards, appraising conformance to these standards, acting when the standards are exceeded, and planning changes in the standards. These activities are conducted throughout the product cycle.

THE PRODUCT CYCLE

Each quality characteristic undergoes an unvarying sequence of activities. This sequence of activities, the product cycle, is diagrammed in Figure 2.1. The cycle begins with the design of the product and the specification of the quality characteristic. After the process required to achieve the specified quality standard is defined, materials are purchased, equipment and procedures are developed, and operators are trained in the required tasks. Then, after inspectors have examined the product to judge its conformance with design, consumers purchase and use the product and this experience is the basis for redesign which starts the cycle all over again.

Quality assurance can contribute to product quality in a number of ways throughout the product cycle. Early in the cycle the product's char-

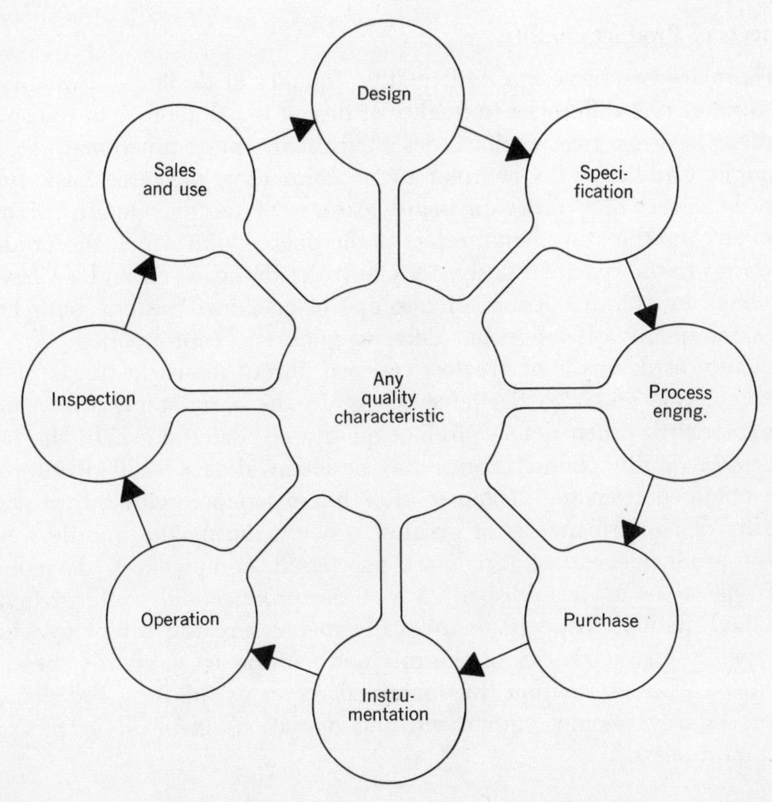

Figure 2.1 The product cycle (adapted from J. M. Juran, *Quality Control Handbook*. New York: McGraw-Hill, 1962.)

acteristics are determined in design activities, quality standards are specified and the manufacturing process is developed. At this time preliminary product designs and process concepts may be reviewed for potential quality problems. As materials and equipment are purchased, procedures may be established for the qualification of suppliers and for the acceptance of purchased items when they are received. In conjunction with production operations product quality is measured and corrective actions are initiated to assure that quality standards are met in the items produced. Finally, information may be obtained from the customer concerning problems associated with the use of the product. This information serves as the basis for investigations and tests to determine the causes of defective products. Corrective actions taken as a result are likely to lead to improved quality characteristics and reduced operating costs.

Aspects of Product Quality

There are two basic kinds of quality—quality of design and quality of conformance. A difference in quality of design is a difference in the specifications between two products designed for the same functional use; for example, Cadillac and Chevrolet automobiles serve the same basic function; however, they differ in many features of design. Quality of conformance, on the other hand, refers to the degree with which the product conforms to the specified design. A Chevrolet that can run and a Chevrolet that cannot run because of improperly machined pistons both have the same quality of design but differ in quality of conformance.

Quality of design is of greatest concern during the early stages of the product cycle and also the later stages. In the early stages the quality characteristics which define product quality are determined. In the later stages the quality characteristics may be changed as a result of information obtained from the customer after his experience with the product. Quality of conformance is of greatest concern during the middle stages of the product cycle. Materials are purchased, equipment is developed, and operations are established so that the product will conform to the specified quality. Inspections are made to measure the extent to which the quality characteristics of the product conform to standards specified for them. The information thus obtained serves as the basis for changes in procedures, people, equipment, and operations involved in the production process.

QUALITY ASSURANCE OBJECTIVES

The overall quality goal of industrial management is to meet the quality needs of specific customers in the most economical manner possible. Although this goal has remained essentially unchanged over the years, the operations involved in obtaining it have not. The trend toward increased complexity and precision of both products and processes has caused an increased emphasis to be placed on quality assurance efforts. It is simply becoming more difficult to meet quality objectives. As a consequence the proportion of product cost typically allocated to quality assurance has increased significantly. Correlated with the increase in quality assurance costs is industrial management's concern with the efficiency of quality assurance operations and ways in which costs can be reduced.

The objective of quality assurance is to obtain the greatest quality value for each quality dollar spent. It is not difficult to see that, as the level of quality assurance activity increases costs will also increase, but, efects in products; engineers solve technical problems; and supervisors lan, decide, and communicate as they organize and direct the activities f others. Should the system break down, it is up to the people to fix it. t is also up to the people, especially the supervisors and engineers, to nprove the system by increasing its capabilities and reducing its costs.

How can good job performance be obtained? The answer to this ques- on is found in the particular set of individual, physical, and organiza- onal factors which influence a person in his job. The general term for iese influences is *human factors*. Individual factors include skill, knowl- dge, attitude, temperament, and interests. Physical factors include work ıyout, tools, equipment, and aids. Organizational factors include work iethods, policies, type of work group, supervisory practices, and the ocial aspects of the organization. These are the factors individually and ı combination that determine how well people do their jobs in a quality ystem. As a consequence of placing a heavy reliance on the human fac- ors in the quality system, many of the problems in quality assurance are uman problems and require the following considerations:

1. Organization of quality assurance tasks into jobs so that the skills nd knowledge required for satisfactory job performance match the skills nd knowledge of available personnel.
2. Development and selection of tools, equipment, and instruments ıat are compatible with and enhance the ability of the people selected) perform quality assurance tasks.
3. Methods of obtaining and providing information to facilitate qual- y decision making and problem solving.
4. Application of techniques to measure performance of personnel for urposes of job improvement, training, and the evaluation of new tools nd techniques.
5. Specification of quality and communication of quality specifications) appropriate people both inside and outside the quality assurance rganization.
6. Selection of people who have the capability for performing given uality assurance jobs and efficiently training these people to an accept- ble level of performance.
7. Development of quality mindedness within people and investiga- on of ways in which quality goals can be matched with the personal oals of individuals.

UMMARY

1. For purposes of quality assurance, quality is defined under specified onditions of product use and product cost.

if quality assurance costs are more than offset by the decrease in costs due to quality losses, quality value is increased. There is a level of quality assurance at which the total costs—cost of quality assurance plus cost of quality losses plus manufacturing costs—is lowest. This is the point at which the value of quality assurance is greatest and, therefore, the appropriate level of quality assurance effort. The relationship between the cost and value of quality assurance is illustrated in Figure 2.2.

The model presented in Figure 2.2 is relatively easy to discuss, but quite difficult to actually apply. The costs of quality assurance and the costs of quality losses are not easily determined. The costs of quality assurance can be determined precisely provided the data are available but these cost data typically exist in many forms and are scattered throughout an organization. Some quality losses such as the costs of scrap, rework, and refused deliveries, may be easy to ascertain; however, other more intangible losses such as the costs resulting from a reduction

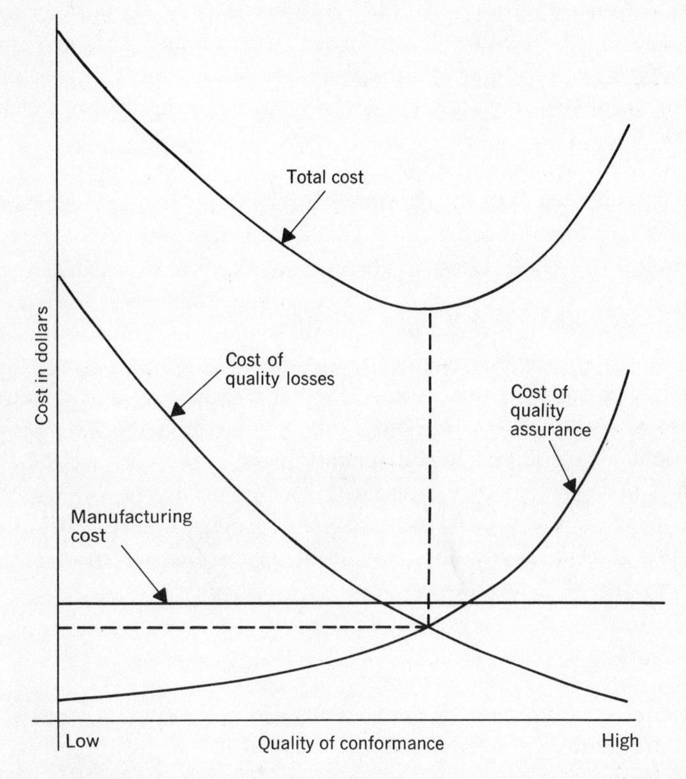

Figure 2.2 Relationship between cost and value of quality assurance (adapted from J. M. Juran, *Quality Control Handbook, loc. cit.*).

in quality image among customers may defy determination in terms of dollars.

Within the framework provided by the basic quality assurance goal more specific quality objectives necessarily reflect the nature of each different enterprise. These specific objectives typically reflect the following aspects of quality assurance.

1. Prevention of losses due to production errors. Increased emphasis is being placed on the role of quality assurance in preventing the error at its source in contrast to detecting the error later on. One approach being taken is the development of quality consciousness among production operators.

2. Efficient measurement of product quality. The measurement of product quality is a major expense of quality assurance. Consequently the amount of information obtained per unit of cost is a relevant concern.

3. Early detection and correction of quality problems. An unsatisfactory production condition can be considered to be a cost generator. Added costs due to losses will continue to accumulate until the unsatisfactory condition is detected and corrected. The amount of quality loss due to the condition is a function of how soon the condition is detected and corrected.

4. Effective communication of quality information. Quality information that is collected but not communicated to the person who can take action is as good as no information at all, and therefore a relevant quality assurance objective is adequate communication of information obtained.

5. Establishment of appropriate quality standards. Standards that are set too low will result in customer dissatisfaction. Standards that are set too high will result in unnecessary costs. An objective of quality assurance is to ensure that quality standards are set at a realistic point for both customer satisfaction and economic production.

6. Assurance that quality standards for materials, equipment, tools, and aids used in the production process are met. The quality of these items, since they influence the production process, are important to the ultimate quality of the product produced.

THE QUALITY SYSTEM

The people, equipment, and information organized to assure that products are made to certain standards of quality can be thought of as a quality system. It is this total system that performs operations directed toward the achievement of specific quality objectives for each set of quality characteristics. The system concept is important in the considera-

tion of quality assurance objectives and activities beca
elements of the system can be considered in isolation. Ea
elements must interact with other elements in the attair
objectives; for example, an inspector cannot detect def
without the information that tells him what quality chara
sider and what the tolerances are with respect to their
In addition, he must use some type of equipment to ma
inspection. Similarly, an item of equipment cannot be co
absence of the person who will be using it or the info
provided to either the person or the equipment.

The quality system engages in four basic types of activ
quality objectives. These are setting quality standards, a
uct conformance to the standards, taking corrective ac
problems, and implementing improvements in the qualit

The establishment of quality standards is a matter of rev
designs in light of customer quality needs and setting the
quality characteristics accordingly. As already discussed,
be considered is between customer needs and the ecor
duction.

The appraisal of the conformance of quality charact
established standards is most often a matter of inspectio
Measurement is the key part of both these activities. The
teristic is measured and compared with acceptable tole
these tolerances are exceeded, the product is rejected an
condition is recorded and communicated so that appropr
action may be taken.

Corrective action may take many forms. It may be as
recting an error in a technical manual or as complex as rec
redesigning the production system. The objective of corre
to prevent the subsequent production of items containing ou
quality characteristics.

Planning for improvement also takes many forms. The
this activity may be a specific problem that has been ide
quality system or simply a general feeling that the qualit
be improved. The typical method used to plan improve
system is to conduct a special study which involves the dev
evaluation of proposed system changes.

HUMAN CONSIDERATIONS IN QUALITY ASSURANC

The worth of a quality system depends largely on how w
their jobs. People are responsible for the success of the syst
the major role in the functions it performs. Inspectors dete

2. The basic element in quality is the quality characteristic. By defining the quality of a product in terms of its composite of quality characteristics a precise definition of product quality can be specified. Acceptable quality, then, is a function of the extent to which the product's quality characteristics meet the standards established in light of customer satisfaction.

3. There are two basic aspects of product quality—quality of design and quality of conformance. Quality of design concerns the degree to which the product is appropriately designed for its specified functional use. Quality of conformance refers to the degree with which the product conforms to the specified design.

4. The objective of quality assurance is to assist industrial management in meeting the quality needs of specific customers in the most economical manner possible. In meeting this objective, quality assurance is involved throughout the product cycle, from the specification of product design to product sales and use.

5. Quality assurance activities should be evaluated in terms of the amount of quality value they produce. Quality value is produced as long as quality assurance costs are more than offset by the decrease in costs due to quality losses.

6. For purposes of improving the effectiveness of quality assurance activities, it is useful to consider the people, equipment, and information organized to conduct these activities as a quality system.

7. The quality system engages in four basic types of activities in meeting quality objectives. These are (a) setting quality standards, (b) appraising product conformance to the standards, (c) taking corrective action on quality problems, and (d) implementing improvements in the quality system.

8. The effectiveness of a quality system in meeting quality objectives depends largely on how well people do their jobs. Job performance is a function of the particular set of individual, physical, and organizational factors that influence a person in his job. The general term for these influences is human factors.

REFERENCES

[1] J. M. Juran, *Quality Control Handbook*. New York: McGraw-Hill, 1962.
[2] A. V. Feigenbaum, *Total Quality Control*. New York: McGraw-Hill, 1961.

III

HUMAN FACTORS RESEARCH AND ENGINEERING

The solutions to human problems in quality assurance are to be found through the methods of behavioral science; however, the solution to a particular problem does not rely on any established method or set of methods. Emphasis is given at the outset to carefully identifying and describing the problem. Human factors research, then, is a matter of developing qualitative and quantitative information about human behavior relevant to the problem. Human factors engineering is the application of this information to the improvement of human performance. The proposed approach therefore is offered as a replacement for the use of "common sense" which has so often led to incorrect and ineffective solutions concerning people in the issues of quality.

POTENTIAL CONTRIBUTIONS

The ultimate contribution of human factors research and engineering is more effective system performance. In a quality system this is greater effectiveness in achieving quality objectives. Since there are several categories of quality objectives, the potential contributions may be discussed more specifically.

A common starting point for a human factors effort is someone's feelings or more objective evidence that the quality system is not performing as well as it should or can be improved with respect to some aspect of human performance. The identification and description of the problem is a major contribution and often the most difficult part of the effort. Once the problem has been identified, the solution is often obvious, requiring only a minimum amount of additional effort. A clear definition of the problem is a large first step forward.

A second potential contribution is in the application of human factors information to the design of the quality system. One of the best ways to ensure effective human performance is to design the system so that the effects of human capabilities are maximized and the effects of human limitations are minimized. There are some tasks that people generally do well and others that they generally do poorly. The idea is to design the system so that people perform only those tasks that they are capable of performing well.

A third potential contribution is the reduction of system costs. A common result of finding a better way of doing something is the reduction of waste. Waste may be associated with any one of the three types of element making up the system—people, equipment, and information. People may be performing tasks that waste their skills; equipment inappropriate to the tasks to be performed may be provided (the result, wasted time and unused equipment), and information inappropriate to the tasks at hand may be gathered (the result, unused information and wasted effort).

The fourth area of potential contribution concerns the development and maintenance of human resources. A reserve of good will and high morale among employees is needed to get the system through difficult times. Almost any system when subjected to a combination of excessive pressures will tend to break down and become less effective. The extra effort required to keep this from happening usually has to come from the people in the system; the elements of equipment and information cannot be readily changed. Consequently human factors efforts which lead to greater job satisfaction and morale among employees are likely to pay great dividends in times of stress.

The fifth and final category of potential contribution is the information fallout from research efforts and the unexpected gains which are obtained incidental to the primary objectives of a study. Information obtained from a research study may be applicable to similar situations which exist currently or which may be encountered in the future. Consequently, additional mileage can be obtained from a set of study results. A frequent experience in research is learning something that one did not specifically seek to find. A recent example of serendipity, a word coined to describe this experience, involved a discovery made during a study of precision measurement techniques. Certain threading gauges were found to add nothing to the effectiveness of machined parts inspectors. Discontinued use of these gauges resulted in an annual cost saving of several thousand dollars.

To highlight the types of improvements that can be made in quality systems through human factors research and engineering, the following examples are listed. These are actual accomplishments that have resulted from the joint efforts of quality supervisors, quality engineers, and human factors specialists.

1. A new method was developed for inspecting complex electronic equipment. It was found to increase inspection accuracy by 90 percent. The new method eliminated much "mental gear shifting" by encouraging inspectors to inspect for one type of quality characteristic at a time.

2. The job of the quality engineer was analyzed and redesigned. As a result, engineers were found to spend 34 percent more time on tasks which required their skill and knowledge and correspondingly less time on tasks which did not. In addition, the job of quality technicians was enlarged to include a number of the tasks previously performed by quality engineers.

3. Techniques were developed to improve the precision measurement performance of machined parts inspectors. Based on an analysis of some initial performance measures, a four-hour, on-the-job training course was designed and a set of visual job aids prepared. Use of training or visual aids alone significantly improved performance; use of both in combination was found to increase inspection accuracy by more than 70 percent.

4. The use of systematically prepared photographs to illustrate quality standards significantly increased agreement among inspectors on borderline quality cases. Use of photographs which defined the borderline between acceptable and unacceptable solder connections, for example, increased agreement among inspectors on borderline cases by 100 percent.

5. Methods developed and employed to increase understanding among supervisors and their work groups were found to greatly improve performance. The following two examples are typical of the results obtained. An inspection group reduced paperwork errors by 75 percent after they set their own goals and received knowledge of performance results on a regular basis. A power supply assembly group reduced its rate of defectives by 40 percent within one month after the initiation of a series of group goal-setting and problem-solving sessions.

LIMITATIONS

Human factors research and engineering efforts do have limitations, the most obvious being their scope. Although other elements in the quality system are considered, these efforts are limited to problems associated with human performance. It is recognized that there are many aspects of quality assurance that are not directly related to problems of human performance.

Human factors studies cost money. Therefore at the time a particular study is contemplated some sort of comparison should be made between the potential gains that may result from the effort and the predicted cost of the effort. Since the resources of money and talent are usually in short supply in industry, all possible research and engineering efforts cannot be undertaken. Consequently priorities for these efforts must be established. It would make more sense, for example, to initiate a study where the potential returns are estimated to be roughly on the order of ten to one than if the maximum gain would only be about two to one. Obviously in either case there is the risk that the gain will be zero; if the outcome were perfectly predictable, there would be no need for a study.

The product of human factors research and engineering is change, a commodity that is often difficult for people to accept. In addition to the technical work involved in a study, effort must be expended toward obtaining the acceptance and implementation of the results. Even if the study results clearly indicate the direction to be taken, it cannot be assumed that the conclusions and recommendations will bring about immediate changes. Implementation of changes likely to be indicated by study results must be planned from the outset. This additional burden is likely to add to the costs of the study and to the constraints under which it is conducted; for example, a relatively high level of participation by those who will be responsible for implementing changes is one of the best ways to incorporate indicated changes. The additional effort required to attain a high level of participation is a necessary part of human factors research and engineering activities.

THE SYSTEM CONCEPT

Human performance should be investigated within the context of the system in which it exists. A common mistake in attempting to solve human problems or in trying to improve human performance is considering people as isolated beings. In doing so the relationships among people and the other elements of the system are ignored. Often the only solutions considered involve an attempt to adapt people to a predetermined set of conditions; frequently the first attempt at a problem solution is one of providing more training or of increasing motivation in some way. These approaches may not necessarily produce the greatest gains in human performance. The system concept provides a basis for considering other critical factors which may lead to improvement in human performance.

The quality system is composed of people, equipment and information.

The system performs operations directed toward the achievement of certain quality objectives for each set of quality characteristics. As shown in Figure 3.1, changes are introduced into the system as a result of human factors research and engineering activities. The changes are designed to improve human performance and thus ultimately to improve the performance of the system in attaining quality objectives. In so doing none of the elements of the system is considered to be permanently fixed. The effects of combinations of changes as well as the effects of changing single elements are considered; for example, the problem of an excessive number of inspection escapes may not necessarily be solved by either disciplining or training inspectors. This kind of solution may fail because it considers only one element in the quality system, namely people, and this element may be the one which provides the least opportunity for improvement. The best solution might be the combination of a new inspection tool, a set of instructions for using it, and a brief period of on-the-job training to quickly develop skills in using the tool.

THE SCIENTIFIC APPROACH

A scientific approach is employed in human factors research to obtain and organize information in a way that will enhance problem solving and decision making. Although it is difficult to define "scientific approach" in a way that would be agreeable to all "scientists" there are several characteristics that distinguish scientific methods from nonscientific methods. These include the objectivity of the investigator, the requirement for measurement, and the exhaustive nature of the investigation.

The objectivity of investigation can be shown by contrasting it with other ways of reducing uncertainty in problem solving and decision mak-

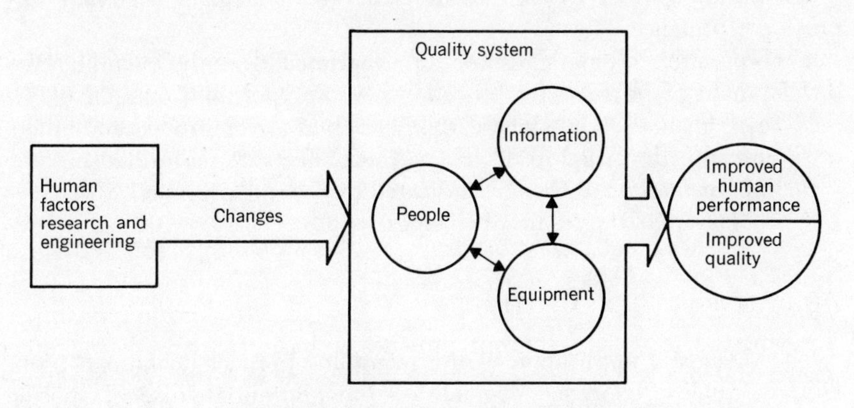

Figure 3.1 The quality system concept.

ing. Objective investigation bases conclusions only on the evidence available and minimizes the extent to which the investigator's own desires and biases enter into the conclusion drawn. By contrast nonscientific approaches include holding a favored opinion regardless of new evidence introduced that conflicts with that opinion, accepting as fact the unsubstantiated opinion of a presumed authority, and concluding that something is "obviously true" in spite of the lack of any supporting evidence.

The scientific method emphasizes accurate measurement. The precision of measurement may vary widely from the precise measurements provided by electronic counters to the relatively crude measurements obtained from questionnaires and rating scales. Any of these measurements is appropriate in a scientific investigation, however, if they are not used for purposes beyond their limitations.

The third characteristic of the scientific method, the exhaustive nature of investigation, concerns the attempt to bring into consideration all of the relevant evidence. Information is not passed over because it fails to fit some previously established pattern. Furthermore, new evidence is sought to support or refute existing conclusions. In this respect the human factors specialist is somewhat of a professional skeptic. He is not easily convinced and he is continually looking for new information that will shed new light on the problems under study. Application of the scientific approach to human problems in quality assurance may take many different forms; however, a typical sequence includes the following seven steps:

1. *Exploration* to define the problem and to identify the types of information that are available and relevant.
2. *Analysis* of available performance data and the collection of new data required to identify the critical factors in human performance.
3. *Development* of hypotheses or ideas about changes that may improve performance.
4. *Preparation* of materials for an experimental study including the development of instructions, procedures, prototypes, and equipment.
5. *Experimentation* in which certain variables are controlled and others are systematically varied to determine their effects on performance.
6. *Implementation* of changes indicated by the study results.
7. *Evaluation* of the results of changes.

HUMAN FACTORS METHODS

The successful application of the scientific approach to human problems in quality assurance depends on the appropriate use of specific research and engineering methods. Within the field of behavioral science there can be found a variety of methods that have been successfully applied in the solution of a variety of human factors problems. It is seldom, however, that a method exists in a form that is ready for direct application to the problem under investigation. Most often previously used methods provide the model for developing new methods that are more appropriate or provide the inspiration for refinements that enable the method to better serve its intended purpose. The variety of specific methods that have been developed and used and that are likely to be developed for future use fit in one of the following four categories: analytical techniques, behavior measurement, experimentation, and application of human factors information.

Analytical Techniques

This category includes the various ways of organizing information for the purpose of making inferences. Included are the many statistical methods used for making probability statements about something that has been observed. Statistical analyses contain three basic elements: a hypothesis, a statistical test, and a level of significance. The statistical hypothesis is a statement that, on the basis of information obtained, one seeks to support or refute. The statistical test is a set of rules whereby a decision about the hypothesis is reached. The level of significance is the probability level at which the hypothesis is to be accepted or rejected on the basis of the statistical test; for example, we may be concerned with the relative effects of tool A and tool B on inspection performance. Having collected performance data from inspection operations in which each of the two tools was used, we are interested in knowing whether the difference in performance for the two tools is sufficiently great to justify our recommending the use of one over the other. Since tool A is currently in use, our hypothesis is that tool B results in better inspection performance than tool A. A statistical test is selected to test the significance of the difference between the average number of defects detected using tool A and the average number detected using tool B. Because we do not want to change tools unless we are very sure that any improvement in performance did not occur by chance, we have established the level of significance at 0.01; that is, the odds against a significant difference at this level having occurred by chance would be 99 to 1. The results of the analysis, then, permit us to make a decision based upon a known probability of our decision being in error.

Another type of analytical technique is that generally referred to as task analysis. The emphasis of task analysis and related types of techniques is on the systematic organization of information. This method is frequently used during the exploration of a problem to assist in defining the problem and identifying the types of information that are available and relevant to the operations under investigation. The basic idea is one of reducing an operation to the basic sequence of activities of which it consists. In the same manner that an X ray exposes parts of the human system that are otherwise not visible a task analysis exposes the parts of an operation that otherwise remain unseen. Once the basic activities of an operation have been identified, a meaningful basis exists for exploring factors which may be critical to human performance; for example, we may find that activity J requires a measurement to be made within a tolerance of 1 mil. Yet we may have evidence that the average error produced by the man-machine system provided for the measurement may be about 5 mils. As a result, we have identified a potential human factors problem in this operation.

Behavior Measurement

Methods of measuring human behavior are many and varied. They include the use of job records, performance tests, questionnaires, electronic error counters, brain wave recorders, interviews, lie detectors, eye cameras, and many more. Obviously, these methods have been developed for a variety of purposes. The use of any given measurement method can be justified only in terms of the purpose for which the information obtained will be used. Many examples of measures developed to obtain information for specific purposes are provided in the last seven chapters of this book. In addition, Chapter VIII discusses in detail the criteria for a good measure of inspection performance and discusses several different performance measures which have been developed.

Experimentation

An experiment may take many forms. It may be as unsophisticated as the attempt to establish a set of optimum conditions by means of a series of trial and error iterations or as sophisticated as a factorial experiment in which a number of variables are simultaneously varied and controlled. Even for the same problem an experiment may be designed in a number of different ways. Winer [1], in his book on experimental designs for research in the behavioral sciences, discusses the role of experimental design.

"The design of an experiment may be compared to an architect's plans for a structure whether it be a giant skyscraper or a modest home. The basic requirements for the structure are given to the architect by the prospective owner. It is the architect's task to fill these basic requirements; yet the architect has ample room for exercising his ingenuity. Several different plans may be drawn up to meet all the basic requirements. Some plans may be more costly than others; given two plans having the same cost, one may offer potential advantages that the second does not."

In designing an experiment, a number of questions must be answered relative to the usefulness of the information to be obtained. Should the experiment be conducted in the artificial environment of the laboratory or in the real environment of the actual work setting? How many subjects are required? What should the characteristics of the subjects be? How should the independent variables be varied? How should the dependent variable or variables be measured? What statistical analysis should be used? The answers to these questions and others will shape the specific nature of the experimentation employed.

Application of Human Factors Information

Although human factors research and engineering is a relatively new field, a great deal of information on human behavior has been developed during the last 25 years; for example, the *Human Factors Engineering Bibliographic Series,* published by the Institute for Psychological Research, Tufts University, lists more than 28,000 publications pertinent to human performance in man-machine systems. Information may exist that is directly relevant to a human factors problem or question that has been identified. The primary effort required may be that of putting existing information into a form that is useful.

An example of one way that human factors information can be applied is the human factors checklist. The checklist is developed from available information on human performance. It consists of a set of statements that can be used in determining the adequacy of a system, operation, item of equipment, set of work instructions, and so on. Thus changes are initiated for system improvement through translating information on human behavior into system requirements.

THE HUMAN FACTORS SPECIALIST

Effective human factors research and engineering in a quality system require a team effort. The team should consist of a human factors specialist in addition to people within the quality organization. Depending on the nature of the effort, the team may include quality supervisors, inspectors, and quality engineers. Some notable success has resulted from the establishment of a human factors coordinator in a larger quality assurance department. A person in this capacity coordinates the requirements for the various continuing efforts in the department, plans assistance to line personnel and acts as a point of contact for administrative matters associated with human factors efforts. This senior staff position should be occupied by a quality engineer with background in statistical methods, quality engineering, and production systems. The human factors coordinator would be a member of the team in each human factors effort conducted within the department.

The human factors specialist brings technical skills and knowledge of industrial and experimental psychology to the team. To be effective in the quality environment the psychologist must also have some training or experience in the physical sciences and in engineering to be comfortable in dealing with problems involving equipment and people who are equipment oriented. He should also have the ability to define research problems from relatively ambiguous situations in addition to possessing a sincere interest in practical quality problems.

To be successful in initiating and conducting a human factors program in quality assurance the human factors specialist must maintain a problem-oriented frame of reference. It is all too easy for the specialist to develop skills in certain human factors methods and rely upon these to the exclusion of others which may be more appropriate. He should recognize that the people he is working with are typically interested in solving their quality problem in the quickest and easiest way possible. As a consequence, the human factors specialist may find himself at odds with those in the quality system. He may wish to conduct a relatively extensive research study that will have benefit not only to some of the immediate problems but to anticipated future problems and problems in other operations. The quality assurance people, on the other hand, may be interested only in a solution that they can apply next week. If the human factors specialist feels that he is correct in this case, he must plan his sales strategy in light of the motivations of those who will not readily understand his program.

SUMMARY

1. The ultimate contribution of human factors research and engineering to the quality system is increased effectiveness. More specific contributions include the identification of human performance problems, the application of human factors information to system design, reduction of system costs, the development and maintenance of human resources, and the increased knowledge about quality assurance operations.

2. The limitations of human factors research and engineering are asso-

ciated with their scope, cost, and acceptance. Although these efforts may have an impact throughout the quality system, they are limited to problems associated with human performance. In addition, their cost should be examined in light of potential gains and the efforts required to obtain acceptance of indicated changes.

3. To avoid the mistake of considering people as isolated beings, human performance should be investigated within the context of the system in which it exists. The system concept provides a basis for considering all of the critical factors that may lead to improvement in the performance of people.

4. The scientific approach is employed to obtain and organize information in a way that will enhance problem solving and decision making with respect to human performance. Characteristics of the scientific approach include the objectivity of the investigator, the requirement for measurement, and the exhaustive nature of the investigation.

5. The solutions to human problems in quality assurance are to be found through the methods of behavioral science; however, the solutions do not rely on any established method or set of methods. The variety of specific methods which have been developed and used and which are likely to be developed for future use fit into the following four categories: analytical techniques, behavior measurement, experimentation, and the application of human factors information.

6. Analytical techniques include the various ways of organizing information for the purpose of making inferences. Included are the many statistical methods used for making probability statements about something that has been observed as well as task analysis that is used to systematically organize information about human activities.

7. Behavior measurement may employ any of a variety of techniques. The use of any given measurement method, however, must be justified in terms of the purpose for which the information obtained will be used.

8. Experimentation is employed to make inferences about treatments or conditions of interest. The design of an experiment may take many forms, depending on the study objectives and the resources available.

9. Although human factors research and engineering is a relatively new field, a great deal of information has been developed pertinent to human performance. This information may be adapted for direct application to human factors problems.

10. The improvement of human performance in quality systems requires a team effort. The team should consist of a human factors specialist in addition to people within the quality organization. It may be desirable to establish a human factors coordinator within a larger quality assurance department.

11. To be successful in improving the performance of quality systems, the human factors specialist must maintain a problem-oriented frame of reference. He cannot be successful by attempting to apply a set of favorite techniques to any problem encountered.

REFERENCES

[1] B. J. Winer, *Statistical Principles in Experimental Design*. New York: McGraw-Hill, 1962.

IV

THE ROLE OF THE SUPERVISOR

Because supervisors interact with the other elements of the quality system —other people, equipment, and information—in an attempt to meet quality objectives, they anticipate problems about people, equipment, and information, and plan to overcome them. Supervisors communicate with other people in the system and make decisions regarding the allocation of their resources. In short, supervision is responsible for an effective operation; if the system is of faulty design, supervision must make changes; if the system breaks down, supervision must fix it.

In the cost conscious, highly competitive environment of modern industry we cannot afford supervisors who overemphasize human relations and see their role as one of making people happy. In an industrial organization whose objectives are growth and increased profits the role of the supervisor is to ensure that his part of the organization meets its specific targets and thus contributes to these overall objectives. In a quality system quality objectives can be effectively met only if the supervisor can mobilize his subordinates and direct their efforts toward specific quality targets. His role is not one of making people happy but one of working through people toward system objectives.

As a result of his position between the people he supervises and company management, the industrial supervisor finds himself the "man-in-the-middle." He is said to be caught between the expectations of management and the wants of his subordinates. If this is true for supervisors in general, it is particularly so for the quality supervisor. In addition to pressures from above and below, he is also subjected to forces from people in production and engineering operations.

What activities should a supervisor engage in and how can he best perform his role in the quality system? A single, unqualified answer does not exist at this time. However, a view of the supervisor's role based on

some of the most recent research and thinking on this subject is discussed in this section; in the remaining sections, techniques and information which should be useful to a supervisor in effectively discharging his responsibilities are discussed in detail.

FUNCTIONS OF SUPERVISION

To be effective in the complex environment of a quality system, the supervisor must have a clear picture of his functions. Adequate job definition is especially important for the new supervisor since he is typically in the process of making the transition from a doer to the first level of management; for example, an inspection supervisor needs to realize that he cannot be effective by being a "superinspector." Since he can typically bring a great amount of inspection skill and knowledge to bear, he will have a tendency to handle difficult inspection problems himself. In general, he will have a tendency to do any task at which he is clearly superior since he sees it as a method of saving time. In a similar manner a new supervisor in a quality engineering organization tends to neglect his supervisory duties and spends a large proportion of his time solving difficult and challenging technical problems. One of the most critical problems in the area of technical supervision results from the fact that these people are generally promoted on the basis of their technical ability. In addition, they usually attain a major portion of the job satisfaction from solving technical problems.

The supervisor who insists on solving all the difficult and interesting problems himself reduces the long-term effectiveness of his organization in two ways: (a) he inadvertently deprives his subordinates of opportunities needed to develop their own abilities and (b) he reduces the time he spends on the activities that are the unique responsibility of supervision. This type of situation degenerates very rapidly to an operation characterized by poor communication and continuous "fire fighting."

One of the hardest lessons a new supervisor must learn is to accomplish things through other people—even when he could do a better job himself. To provide a basis for developing effective supervision, the general functions of management are reviewed below. Independent of the assumptions under which he operates, there are certain functions that must be performed by every supervisor to meet his objectives. These functions can be categorized roughly as planning, decision making, communicating, monitoring performance, and personnel development. The effectiveness of a supervisor depends to a large extent on his skill and knowledge in these five basic areas.

Planning

The success of any quality operation will depend on the supervisor's ability to anticipate future needs, establish challenging goals for his operation and provide the resources needed to reach these goals. Examples of some specific planning activities which are required of inspection and quality engineering supervisors are the following:

1. Forecasting the inspection personnel requirements for future programs.
2. Scheduling familiarization and training programs to prepare quality engineers for future assignments.
3. Forecasting inspection equipment and task requirements for a new product.
4. Requesting studies to provide timely information for the design of new jobs.
5. Setting short-term and long-term goals with employees.
6. Budgeting personnel and equipment resources to reach quality goals in a cost effective manner.
7. Estimating quality information requirements for future programs.

Decision Making

At the core of the quality supervisor's job is the decision-making function. Broadly conceived, decision making entails a collection of appropriate information, organizing the information into meaningful categories, and selecting a specific course of action. Decision making is not unique to the supervisor's job; inspectors, for example, typically make thousands of decisions each day. A crucial aspect of the supervisory decision-making process is the context in which the decisions are made. Supervisory decisions usually involve the management of resources—people, equipment, and information—in the quality system to increase its effectiveness in achieving quality goals.

One of the most difficult tasks in management is that of obtaining adequate data for critical decisions. It is essential that the quality supervisor recognize this problem and assume complete responsibility for assuring that his decisions are based on the best available data at the time when a decision must be made. He should not hesitate to question the adequacy of information at hand and should take the initiative in developing better measurement techniques when they appear to be needed. Performance measurement methods are discussed in Chapter VIII. Measurement is a critical part of the decision-making process because experience and general observation do not provide an adequate basis for selecting the appropriate course of action in many situations; for example, the decision regarding the most effective type of inspector training, the appropriate level of magnification, or the effectiveness of alternative selection techniques all require analysis of inspection performance data.

Communicating

To obtain the information needed for planning and decision making, the supervisor must communicate with other individuals. To implement the results of his planning and decision making, the supervisor must again communicate with others. In the process of this communication he secures an understanding between himself and other individuals. It is important to emphasize that communication is *not* simply talking to someone or sending him a piece of paper. Communication takes place only when an understanding is reached which leads to the desired action. As a result, effective communication requires a two-way transmission of information. Without two-way communication there is no confirmation that an understanding has been reached. In addition, effective communication generally requires follow-up to see that the understanding has resulted in the desired action.

The methods available for transferring information can be easily categorized as written and oral communications. Written communications include letters, bulletin boards, newspapers, posters, suggestions, reports, manuals, books, and questionnaires. Oral communications include person-to-person contacts of many kinds such as conferences, training sessions, telephone calls, and public-address systems. The primary advantages of written communications are authority, accuracy, permanence, and the coverage that can be obtained through their use. The primary advantages of oral communications are flexibility, provision for immediate two-way exchange, and personal effectiveness.

Since written and oral communications each have unique advantages, it would seem reasonable to expect that a combination of the two methods would provide the best results. This expectation has been supported by the results of a study made on the effectiveness of five different communication methods [1]. The study was conducted in various departments of an industrial organization. Tests were given to determine the amount of retention of information transmitted by written, oral, combined written and oral, bulletin board, and grapevine methods. Results indicated that the combined use of written and oral methods was superior to either of the two alone. These results are presented in Table 4.1. Unfortunately,

Table 4.1 Relative Effectiveness of Five Communication Methods

Method	Number of Employees Tested	Average Test Score*
Combined written and oral	102	7.70
Oral only	94	6.17
Written only	109	4.91
Bulletin board	115	3.72
Grapevine only	108	3.56

* All values except the last two differ statistically at the 0.05 level of significance.

many supervisors place communications at a low level of importance relative to their various activities. Often, the need to communicate is regarded as a nuisance and little effort is put forth to do a good job of relaying information and documenting accomplishments. As a consequence, messages are distorted, lacking in clarity, or altogether nonexistent. The system breaks down because of differences in understanding with respect to the objectives and methods and the lack of fundamental knowledge about the condition of the system. If the supervisor does not view communications as a very important part of his role then his efforts in effective planning and decision making will have little if any impact on the quality system.

Monitoring Performance

To assess the effectiveness of planning, decision making, and communicating, the supervisor must have accurate methods for monitoring the performance of his employees. One of the first things a new supervisor must learn is to shift his thinking from how a job should be done to defining job objectives so that the end product can be measured or defined in some meaningful way. The more responsibility a supervisor has, the more impossible it becomes to control, in detail, the way that people spend their time. As a result, it becomes more and more important to develop adequate ways for monitoring results. Because of its recognized importance in inspection operations, a complete chapter has been devoted to the topic of inspection performance measurement. Some guidelines for the development and use of performance measures are as follows:

1. The end product should be defined in sufficient detail so that both the supervisor and the employee can agree on the work to be performed.
2. The quality and time standards for a given job should be clearly communicated before the employee starts the job.
3. The method of measurement should allow for rapid feedback to both the supervisor and the employee.
4. Both positive and negative results should be reported so that the employee receives credit for accomplishments as well as information on how to improve his performance.
5. The performance measurement system should provide the supervisor with information on the problem areas which may require his attention.
6. The performance measurement system should be reviewed periodically to eliminate information which is not required by either the employee or his supervision.

Personnel Development

A planned program of personnel development should start on the first day that each employee reports to the job. The supervisor's first objective should be to set the employee at ease and reduce the anxiety which he may feel about functioning in a new position. This can generally be accomplished by encouraging the employee to talk about himself and providing him with accurate information about his new responsibilities. Information should also be provided concerning the company's general policies and the specific rules that affect his conduct should be explained in detail. Particular emphasis should be given to the safety regulations and the reasons behind them. In addition, attempts should be made to show the worker how his job fits into the department's end product and company goals. Throughout this orientation the supervisor should keep in mind his two main objectives: (a) to provide the employee with a clear picture of his new job and its relation to the company's objectives and (b) to set the employee at ease and reduce the fears that are inevitably associated with a new position.

Once the employee is completely integrated into the work group and performing his initial assignment with a high degree of competence, the supervisor should begin to expand the individual's capabilities by providing opportunities for learning new tasks. As the process of personnel development continues, the supervisor will be required to assign some tasks to his subordinates for training purposes even though it would be easier and quicker for him to complete the tasks himself. It is recognized that many supervisors feel that they are too busy for this type of individual attention to the personnel development function. Personnel training, however, is too important to the continuing success of an organization to be left to chance or scheduled at slack periods of time. As a result of its importance a complete chapter in this book is devoted to the training function in the quality assurance organization. The purpose of this chapter is to describe methods that can be employed by the supervisor to improve the performance of quality assurance operations.

One of the supervisor's most critical duties in the area of personnel development is the identification and training of individuals with a high degree of supervisory potential. These individuals must be identified early in their career and gradually given increased responsibility for planning and decision-making activities. In addition, their skills in the communication and leadership area should be developed through supplementary training. Unfortunately, many supervisors tend to neglect the training of their replacement and some even tend to retard the development of capable individuals who threaten their leadership. Contrary to the fears of many supervisors, they are much more likely to be pushed up than pushed out by competent men under them. In a very real sense the supervisor has not completed his job until he is no longer needed in that position.

When a supervisor has mastered the above functions and learned to teach them to other individuals then he will have the time required to motivate his people and function as an effective group leader.

MANAGEMENT STYLE

Most people would agree that the supervisor's job includes planning, decision making, communicating, monitoring performance, and personnel development, but supervisors differ greatly in the emphasis they place on each of these functions and the specific techniques they use to accomplish their job objectives. These differences in supervisory style are based, in part, on differences in the basic assumptions that are made about human behavior. To be effective in performing the functions of supervision, it is important to be sure that our assumptions about people are accurate.

Assumptions About People

Whether he recognizes it or not, every supervisor operates on the basis of certain assumptions of human behavior. These *assumptions,* because they influence almost every act of supervision, can have significant impact on the effectiveness of the supervisor and, in turn, on the effectiveness of the quality system. McGregor [2] has summarized a number of typical assumptions about people into two categories which have become known as theory X and theory Y.

Theory X is a view of human behavior in industry as characterized by the following assumptions:

1. Most people have an inherent dislike of work and will avoid it if they can.
2. Most people require close supervision and direction through the use of persuasion, control, rewards, and punishment to work towards organizational goals.
3. Most people prefer to be directed, avoid responsibility, and want security above everything else.

This rather negative view of human behavior leads to supervisors who specify rigid standards for work and establish stringent rules and regulations that are vigorously enforced. The emphasis of theory X is on *control* of work behavior through exercise of authority.

Theory Y, on the other hand, is based on the idea that conditions can be created in which people best achieve their own goals by directing their effort toward the achievement of the goals of the organization. Theory Y assumes the following:

1. Most people like to work; the expenditure of physical and mental effort in work is as natural as play or rest.

2. Under proper conditions, most people will exercise self-direction and self-control to attain goals to which they are committed.
3. In modern industry the intellectual potential of most people is only partly utilized.

It is the responsibility of supervision to find new ways of organizing and directing human effort to capitalize on employee potential.

It is very difficult to prove the validity of either set of assumptions since an organization tends to produce people who are consistent with its theory of management. A system set up to control people as though they were irresponsible, undependable, and indifferent is likely to encourage them to behave as such. On the other hand, if people are treated with self-respect and personal dignity, in line with theory Y, they will be encouraged to behave in a manner that is consistent with this positive view of human behavior.

It would appear that a supervisor has a choice between the two approaches in dealing with his employees. He can take the negative view of theory X and by acting in accordance with its assumptions become convinced that people do require a higher degree of control. Probably the more control he exercises, the more will be required. There is, however, an increasing amount of evidence that the supervisor who applies some version of theory Y will experience better performance over the long term. Theory Y has also gained a great deal of acceptance because it tends to result in greater job satisfaction for both the supervisor and his employees.

In spite of the increasing acceptance of theory Y, any experienced supervisor can recall a number of people who clearly met the assumptions of theory X. In fact, most supervisors can expect to encounter some groups that would respond more appropriately to theory X. Because of these individual differences, the typical supervisor cannot rely on a general theory of management. He must learn to provide *flexible leadership* which is appropriate for each group and each situation.

Dimensions of Leadership

Numerous studies have been conducted to determine the style of leadership which produces the best results. In an office situation the supervisors of the most effective sections tended to be employee centered and democratic [3]. On the other hand, supervisors in the less effective office situations tended to be production centered and authoritarian. In spite of these trends, it would be easy to find a number of situations, and perhaps whole industries, where production-centered, authoritarian supervision would be much more effective.

Considering all the differences in people, supervisors, and jobs, it still has been possible to identify two basic dimensions of leadership behavior that have a great deal of generality. These dimensions were based on employee descriptions of supervisory behavior for a large number of jobs and situations [4]. The two factors, *initiating structure* and *consideration*, are independent and should not be considered as opposites. A supervisor may be high on both, low on both, or anywhere in between. Initiating structure involves the planning, measuring results, and feedback functions which were discussed above. A supervisor who is high on this dimension, defines each job in detail, plans ahead, establishes specific methods for each task, and emphasizes group output. Initiating structure is essentially the definition and assignment of work so as to achieve the organization's goals. The second dimension, consideration, deals with the extent to which employees have a sense of mutual trust and respect for their supervisor. Supervisors who are high on consideration are sincerely concerned with the needs of their employees and they encourage open, honest, two-way communication.

Most effective supervisors tend to be above average on both dimensions. The differences, however, between various work groups in the optimum level for each dimension are marked. These findings suggest that the effectiveness of a supervisor may be more dependent on his flexibility than his rating on any predetermined dimensions of leadership; for example, a supervisor might employ a high degree of consideration with very little structure if he wanted a group of experienced assembly people to come up with more effective production techniques. On the other hand, the same supervisor might use a high degree of structure with a minimum amount of consideration for a group of new employees with several known troublemakers.

The basic point is that a good supervisor should be able to identify the critical factors in a job situation and adjust his leadership style to meet the needs of his particular group. However, he must accept the fact that his job requires accomplishing things through other people. To this end it is frequently more effective to obtain cooperation and whole-hearted support for job objectives than it is to rely on detailed direction and control.

Participative Supervision

Management techniques which can stimulate continuous improvement in industrial operations are required to maintain a competitive position in modern industry. Group participation is one form of modern management which has been used by a number of organizations to obtain deeper commitments to company goals and improve employee performance [5]. The implementation of group problem-solving techniques usually requires some major changes in the behavior and attitudes of management, supervision, and hourly employees.

Management must spend a larger share of its time defining specific goals, measuring progress towards these goals, and providing feedback to the supervisor and his employees. In addition, emphasis on detailed control of job methods must be minimized and work groups must be encouraged to participate in decisions which influence their effectiveness. The supervisor must develop skill in obtaining group commitments to organization goals and acquire competence in conducting group problem-solving sessions. The extent to which the supervisor takes action and provides rapid feedback on problems and solutions that affect his people is a critical factor in the success of participative techniques. Finally, the employees themselves must develop a high level of confidence in their supervisor and management so that they will be willing to provide accurate information on the group's performance and participate in open discussions on critical problem areas.

Contrary to the initial reactions of many manufacturing and quality people, the use of group participation results in stronger supervision than traditional forms of management which emphasize detailed direction and control. It forces the supervisor to establish open communications with his people and provides detailed information on the areas of potential job improvement which is seldom obtained by formal information systems. In addition, the group should be able to solve a number of the problems that are typically "delegated to management." As a result, the supervisor should have more time to spend on the legitimate management functions.

Perhaps the most important factor in teaching a supervisor to use participative techniques effectively is the type of supervision they receive. Individuals generally use their supervisor's behavior as a model and assume that the same is expected of them. This implies that the best way to train good supervisors is for management to adopt the role it wishes to have accepted and then reward the supervisor when he displays the desired behavior. In fact, this may be the only way of implementing participative management; research has shown that formal supervisory training has little or no effect on subsequent job behavior when consistent management techniques are not practiced and rewarded by upper management [6].

The potential value of group participation can be illustrated by the following application [7]. A new, first line production supervisor was given eight hours of training to develop the skills, knowledge, and attitudes required to work effectively with small employee groups. Group discussions, on-the-job applications, and individual coaching were provided in an attempt to develop the supervisor's confidence in participative management. As part of the training the supervisor conducted meetings with a group of 11 electronics assemblers.

During the initial series of four meetings, the problems which limited production were discussed and goals were established for quality and output. Based on an average level of 1.5 defects per board, the group established a goal of 0.5 defect per board. The average defect level two months after goal setting was 0.6 defect per board. These results are shown in Figure 4.1.

Based on a stable output of 25 boards per week, the supervisor and his group established an output goal of 50 acceptable boards per week. It is significant to note that this level of output was requested by the supervisor to meet schedule commitments; this goal was not accepted by all the

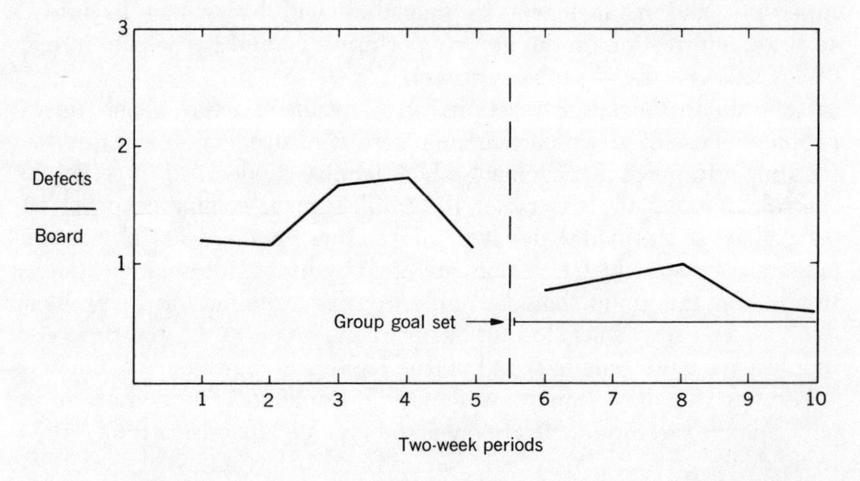

Figure 4.1 Quality performance for a circuit board assembly group.

group members until after a number of production problems were discussed and potential solutions were provided by the supervisor.

Nine weeks after the goal was established the group reached an average of 40 boards per week; this level was maintained for two months. After five months the group finally achieved its goal of 50 boards per week; at this time the requirement for this group of boards was reduced by 50 percent and new work was assigned to this group. The reduction in assembly time for this improvement is shown in Figure 4.2.

In terms of production cost the average assembly time per system (five boards) was reduced from 110 to 40 hours. This resulted in a cost savings of more than $25,000 during the six-month period following the problem-solving/goal-setting process. Critical factors in this improvement were improved communication, increased commitments to production goals, and implementation of employee ideas for better tools and techniques. Additional information on the successful use of group participation is provided in the chapter on employee motivation.

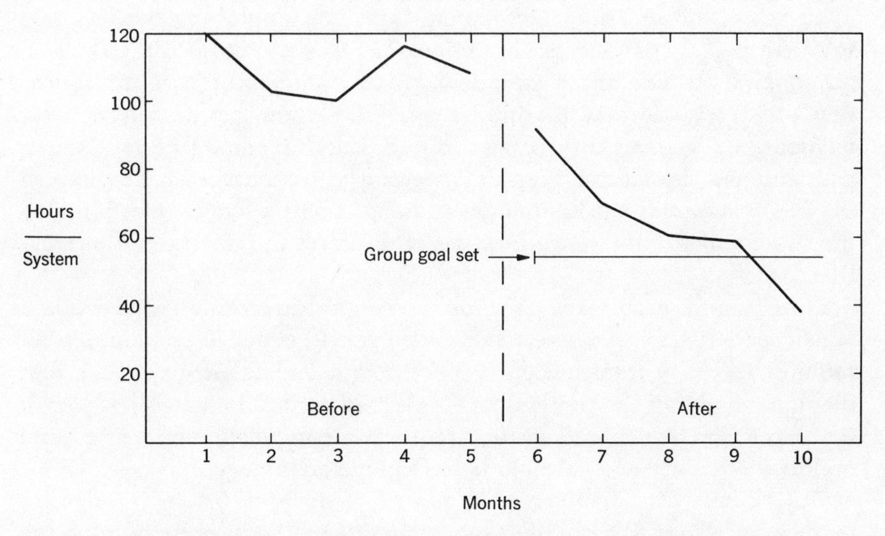

Figure 4.2 Cost reduction for a circuit board assembly group.

SUMMARY

1. Modern industry cannot afford permissive supervisors who overemphasize human relations and see their role primarily as one of making people happy. The effective supervisor must learn to contribute to the organization's objectives of growth and increased profit by working through other people.

2. The supervisor contributes to the organization's goals by performing the functions of planning, decision making, communicating, monitoring performance, and personnel development. One of the hardest lessons for a new supervisor to learn is to accomplish things through other people—even though he feels he could do a better job himself. When he yields to this temptation he deprives his subordinates of opportunities to develop and he reduces the time spent on the functions which are his unique responsibility.

3. The planning function includes the following specific activities: (a) forecasting future personnel requirements, (b) scheduling and conduct-

ing on-the-job training to prepare personnel for future assignments, (c) predicting equipment and facility requirements, (d) requesting appropriate studies to provide planning information, (e) establishing goals and obtaining employee commitments to these goals, and (f) budgeting personnel and equipment resources to meet goals in a cost effective manner.

4. The decision-making function includes collecting appropriate information, organizing the information into meaningful categories, and selecting the specific course of action. The supervisor should assume a questioning attitude about the adequacy of traditional forms of information and he should take the initiative in developing better measurement techniques whenever they appear to be needed. Human factors research has provided the information for increasing the accuracy of decisions in the following areas: (a) employee training, (b) personnel selection, (c) the development of new work methods, and (d) the selection of alternative tools.

5. Communication takes place only when an understanding is reached which leads to the desired action. As a result effective communication requires two-way transmission of information and follow-up to see that the understanding has resulted in the desired action. Research has shown that most effective communication requires a combination of written and oral methods since each technique has unique advantages.

6. To assess his effectiveness in planning, decision making, and communicating, the supervisor must have accurate methods for monitoring the performance of his employees. An adequate system of performance measurement should include: (a) a definition of the end product in terms of quality and time requirements, (b) rapid feedback to the supervisor and the employee of both positive and negative results, (c) information of problem areas which may require the supervisor's attention, and (d) a provision for periodically updating the measurement procedure.

7. The supervisor's responsibilities in personnel development include orientation of new employees, informal training to provide skills and knowledge needed for current assignments, and the identification and training of individuals with a high degree of supervisory potential. In meeting his responsibilities for personnel development the supervisor will be required to assign some tasks to his subordinates for training purposes even though it would be easier and quicker for him to complete the tasks himself.

8. Since the supervisor must accomplish his job through other people, his effectiveness is influenced to a large extent by the accuracy of his assumptions about human behavior. In general, effective supervisors tend to assume that conditions can be created in which people will best achieve their own goals by directing their effort toward the goals of the organiza-

tion. They also recognize the fact that the intellectual potential of most people is only partly utilized in modern industry and they accept responsibility for finding new ways of realizing employee potential.

9. Research has shown that the most effective supervisors in office situations tend to be employee centered and democratic. Less effective supervisors in the same situation tend to be production centered and authoritarian.

10. Although research provides some general guidelines for effective leadership, the highly effective supervisor must learn to provide flexible leadership which is appropriate for each group and each situation. He must be able to identify the critical factors in each job situation and adjust his method of dealing with people to the needs of this particular group. Although the supervisor should not be overly permissive or human relations oriented, it is frequently more effective to obtain the cooperation and whole-hearted support for job objectives than it is to rely on detailed direction and control.

11. Group participation is one form of modern management that has been used by a number of organizations to obtain deeper commitments to company goals and improve employee performance. The implementation of such techniques as group problem solving and goal setting usually requires some major changes in the behavior and attitudes of management, supervision, and hourly employees. The potential value of participative techniques is illustrated by an application which produced highly significant improvements in quality and group output. In a group of electronic assemblers the defect rate was reduced by more than 50 percent and average assembly time per unit was reduced from 110 to 40 hours. This performance improvement resulted in cost savings of more than $25,000 during the six-month period following the group problem-solving/goal-setting process. Critical factors in this improvement were (a) improved communications, (b) increased commitment to production goals, and (c) implementation of employee ideas for better tools and techniques.

REFERENCES

[1] T. L. Dahle, Transmitting Information to Employees: A Study of Five Methods. *Personnel,* 1954, **31,** 243–246.

[2] Douglas McGregor, *The Human Side of Enterprise.* New York: McGraw-Hill, 1960.

[3] D. Katz, N. MacCoby, and N. C. Morse, *Productivity, Supervision and Morale in an Office Situation.* Survey Reseach Center, Institute for Social Research, University of Michigan, December 1950.

[4] E. A. Fleishman and E. F. Harris, Patterns of Leadership Behavior Related to Employee Grievances and Turnover. *Personnel Psychol.,* **15,** 1962, 43–56.

[5] M. S. Viteles, *Motivation and Morale in Industry.* New York: W. W. Norton, 1953. P. 164.

[6] E. A. Fleishman, Leadership Climate, Human Relations and Supervisory Behavior. *Personnel Psychol.*, 1955, 6, 205–222.

[7] F. B. Chaney, Personnel Training and Evaluation for Quality Motivation. In *Quality Motivation Workbook.* Milwaukee: American Society for Quality Control, 1967.

V

DESIGNING THE QUALITY ENGINEERING JOB

Within a typical quality assurance organization, the quality engineer is responsible for a wide range of activities which influence product quality and cost. He may establish quality goals, provide information to inspection personnel, monitor quality levels, investigate problems, initiate corrective actions, and report the results of his various activities. Effective quality engineering requires people with specific skills and knowledge and requires that the engineer's job be designed to take full advantage of his background. In addition, effective quality engineering requires people with the motivation to use their specific skills and knowledge. The job of the quality engineer and techniques that may help in enhancing the effectiveness of people performing this job are discussed in this section.

QUALITY ENGINEERING FUNCTIONS

The specialized functions performed by a quality engineer will depend on a number of things, including the particular process with which he is concerned, specific directions given to him by supervision, and the nature of the quality system within which he is operating. It is possible, however, to establish a list of functional categories and objectives which are generally applicable to quality engineering operations whether they are performed in an aerospace company or in the garment industry. A list of quality engineering functions and objectives was developed by reviewing the reported quality assurance operations in a variety of different types of companies and by an intensive investigation of the activities of a sample of quality engineers in one company. The company quality operations analyzed were those reported at the annual meetings of the American

Society for Quality Control over the past several years. The intensive investigation of a sample of quality engineers was conducted at the Autonetics Division of the North American Rockwell Corporation.

A variety of techniques was employed to obtain detailed task information from a sample of quality engineers studied. Engineers and their supervisors were interviewed at length about the functions they performed and the tasks in which they engaged. On-the-job observations were made by the study team and individual on-the-job diaries were kept by selected engineers. From these data a list of the tasks performed by quality engineers was developed. These tasks were placed into major functional groupings to provide a general statement of the type of work performed. The major functions and their objectives are shown in Table 5.1. A discussion of each of the major quality engineering functions is provided below.

Table 5.1 Basic Quality Control Functions and Objectives for Task Skill Level Classifications

Functions	Objectives
Quality planning	To establish quality goals, adequate acceptance criteria, and inspection equipment requirements.
Providing inspection information	To provide adequate inspection information including visual aids, checklists, and sampling plans; to assure that adequate training is provided for inspecting new products and processes.
Quality monitoring	To collect, analyze, and interpret information on quality performance including data on the accuracy of the manufacturing process and product inspection.
Problem solving	To identify, investigate, and recommend solutions to specific problems.
Corrective action	To obtain the specific changes required to establish and maintain acceptable quality levels.
Documentation	To provide adequate records of activities, quality problems, and corrective actions.

Quality Planning

Coordination and review activities play a significant role in quality planning. Coordination is required with other organizations within the company to assure that quality goals and acceptance criteria are established in light of the relevant information available. Reviews are conducted of proposed designs and process specifications in anticipation of quality problems. The primary planning tasks performed by quality engineers include: (a) establishing quality goals for manufacturing work centers in coordination with manufacturing supervision, (b) representing the quality assurance organization in meetings with other functional organizations, such as engineering, marketing, and manufacturing, regarding the establishment of acceptance criteria, (c) analyzing inspection equipment needs and providing recommendations for the acquisition of inspection equipment and tooling, (d) reviewing proposed general process specifications to assure that adequate quality assurance provisions exist and coordinating revisions with other organizations, and (e) conducting design reviews on new products or processes and indicating potential problem areas to affected quality assurance elements and quality engineers.

Providing Inspection Information

A primary function of quality engineering is determining inspection information requirements and fulfilling these requirements through the development of information sources such as instruction documents, visual aids, checklists, and sampling plans. In addition, quality engineering assesses the training needs of inspectors and aids in the satisfaction of these needs in the most efficient ways available. Activities performed by the quality engineer in carrying out this function include conducting analyses which highlight inspection information needs, making technical recommendations with respect to alternative inspection methods, reviewing and approving information sources such as defect classification matrices, sampling plans, visual aids, instruction documents and checklists, and evaluating inspection operations for changes in information sources.

Quality Monitoring

Just as a radar operator in a military system observes his radar scope for evidence of the enemy, a quality engineer monitors his sources of information to detect the presence of quality problems. However, the radar operator has essentially a single source of information, but the quality engineer must collect, analyze, and interpret information from a number of sources. In addition, the quality engineer is concerned with both the manufacturing process and the inspection process; that is, he must be as aware of problems inherent in his information sources as in the ultimate problem of the manufacturing process. Quality monitoring activities include reviewing and interpreting information indicative of quality problems, conducting analyses of inspection data to obtain detailed information on critical operations, reviewing information on manufacturing processes to assure that economic control of quality is being maintained, and reviewing information obtained from audits of the inspection process.

Problem Solving

The quality monitoring function leads directly to the problem-solving function. As he collects and analyzes information on quality performance, the quality engineer is looking for potential quality problems. These data may indicate that a problem is present but they usually do not identify the specific nature of the problem. Consequently they do not lead directly to the solution or to the corrective action that must be taken. The problem-solving function therefore is a matter of initiating an investigation of the process to determine the specific aspects of the process that are in difficulty. When the problem has been identified and described, the quality engineer considers solution alternatives. At this point the solution may actually be quite obvious, in which case, the problem solution is decided on and corrective action is initiated immediately. In other cases there may be several equally attractive alternatives and there may be additional considerations such as the timing required to initiate the solution, the costs of the solution, and the temporary or permanent nature of the solution. These considerations may require additional study and evaluation by the engineer. If the problem is heavily weighted with human considerations, further investigation of the problem may be conducted in conjunction with a human factors specialist. This additional investigation would be undertaken to provide more and better information that can be used in evaluating the alternative solutions.

Corrective Action

Corrective action refers to those steps taken to obtain the specific changes required to establish and maintain acceptable quality levels. As mentioned in the preceding paragraph, corrective action may follow almost immediately from the identification of the problem. When the problem involves a number of considerations, however, corrective action may not be taken until a relatively exhaustive study is made of alternative problem solutions. In initiating corrective action the quality engineer may be required to coordinate his recommendations among other concerned organizations, to develop and describe the specific actions that must be taken to reach a solution to the problem, to defend his recommended solution at meetings of interested parties, and to review and comment upon critiques of his proposed solutions.

Documentation

As it is with many other activities conducted in industrial organizations, the quality engineer must document his activities. The two basic types of report which are his responsibility are those he prepares of his activities for his immediate supervision and the reports he prepares for interested parties relative to his investigations and the actions he has taken on special problems. His activity or status report is provided to let his supervision know how he stands on his current assignments. His special reports are prepared to communicate to other concerned parties the nature and results of his investigations and his conclusions or recommendations based upon them.

JOB DESIGN TECHNIQUES

One approach to increasing the effectiveness of technical and professional personnel is improving the design of their jobs. The basic idea in structuring engineering jobs is to have the persons who fill them spend a maximum amount of time on tasks that require their skill and knowledge and to reduce a minimum their time on tasks that do not. Adequate job design also considers informational requirements and motivational values associated with the tasks that make up the job. The general subject of job design and the approaches involved are discussed elsewhere [1, 2]. Job design techniques which are applicable to the functions performed by quality engineers such as rating techniques to determine the skills required for each quality engineering task and the motivational value of each task are described in the following paragraphs.

Engineering Task Analysis

Designing the quality engineering job consists of determining the specific tasks which are to be performed by quality engineers in meeting their assigned objectives. Task information may be obtained from a variety of sources including interviews with quality engineers and their supervisors, observations of engineers as they perform their quality functions, diaries which may be kept on the job, and reference books concerning quality engineering functions. After compiling information from these various sources into a master list of quality engineering tasks, the list should be presented to a representative sample of engineers and supervisors. The list may be modified to obtain agreement among these persons. The end product of the task analysis is a detailed task description of the quality engineering job, a description that is agreeable to all parties immediately concerned. This task description will serve as the starting point for using the available skill and knowledge to best advantage and for keeping motivation in the quality engineering job at a high level.

Task Skill Measurement

A skill-level scale may be developed and applied to each task in the master list. A sample scale is provided in Table 5.2. Each task is typed on separate 3 by 5 inch card and rated independently by each of a sample of persons including engineering supervisors, engineers, and the job designers. The deck of task cards is reshuffled for each judge so that the order of presentation cannot systematically influence the skill level ratings. In this manner skill ratings for each task are determined.

Table 5.2 Task Skill-Level Scale

Level	Definition
A	Requires the skills and knowledge of a quality engineer.
B	Can be adequately performed by a quality technician under the close direction of a quality engineer.
C	Can be adequately performed by a quality technician under little or no supervision.
D	Can be adequately performed by a clerk under the direction of a quality engineer.
E	Not a legitimate quality engineering task.

Task Incentive Measurement

To arrive at the amount of incentive each quality engineering task provides, each task is rated on a scale of task satisfaction. An example of a rating scale developed for this purpose is shown in Figure 5.1. In one study of quality engineering task incentives eight different motivational characteristics were investigated using scales of this type.

Importance	Satisfaction
Interest	Liking
Identification	Recognition
Challenge	Advancement

In this study task ratings for these eight characteristics were intercorrelated. It was found that the correlations among these eight characteristics were very high. It was concluded therefore that just one of the more general characteristics, task satisfaction for example, could be used just as well as using all eight. By use of the one characteristic the amount of time required to produce a reliable index of task incentive would be significantly reduced.

Measurement of Task Time Allocation

Estimating the relative amount of time spent on each task by a typical quality engineer can be accomplished by giving a task list to each engineer in a sample group and having him allot 100 hours among the different tasks in accordance with how he spent his time during the previous

A TASK MAY BE QUITE FRUSTRATING AND UNPLEASANT WHILE YOU ARE DOING IT AND STILL GIVE YOU A GREAT DEAL OF SATISFACTION WHEN COMPLETED. HOW MUCH PERSONAL SATISFACTION DO YOU OBTAIN BY COMPLETING EACH TASK? PLEASE PLACE THE NUMBER FOR EACH TASK BELOW THE APPROPRIATE SCALE VALUE.

BEFORE MAKING EACH JUDGMENT, READ THE STATEMENT ON THE CARD AND ASK YOURSELF - "HOW MUCH PERSONAL SATISFACTION DO I OBTAIN FROM COMPLETING THIS TASK?"

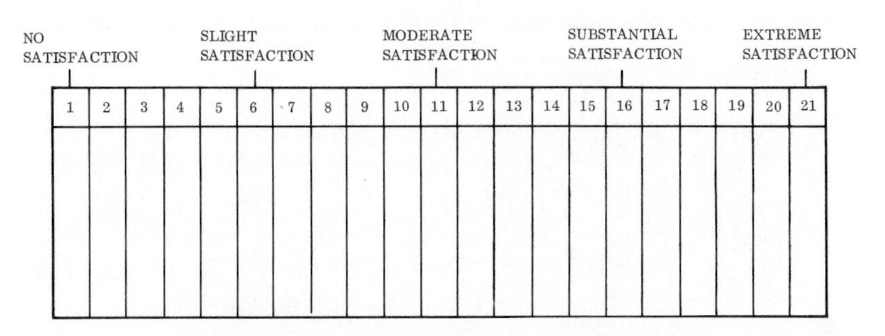

Figure 5.1 Scale for rating task satisfaction.

month. Both the reliability and validity of this procedure have been investigated. In one study each of six engineers completed two time allocation estimates, one for each of two successive months. There was a passage of about a month's time between the two sets of estimates. The rank order correlation between the two sets was relatively high (0.83), and indicated a high level of agreement between the two estimates. The validity of the estimates was tested by correlating the estimated allocations with allocations obtained from daily task time logs maintained over a two-week period. This correlation was sufficiently high (0.74) to indicate that roughly the same information could be obtained from time allocation estimates as from the more burdensome daily task time logs.

QUALITY ENGINEERING APPLICATION

Job design techniques were applied to activities performed by a sample of quality engineers responsible for quality objectives in the production of electronic, electromechanical, and mechanical devices. The objective was to redesign the existing job so that more effective use could be obtained of quality engineering skills and knowledge. This application demonstrates the gains that can be made through the systematic design of quality engineering jobs.

Task Skill Level Versus Time Allocation

To determine how effectively the skills of engineers were being used in quality engineering activities under the existing job design, an analysis was made of the percentage of time expended on engineering-level tasks as opposed to the percentage of time expended on technician-level tasks. Task data indicated that engineers were spending only 51 percent of their time on engineering-level tasks; the remainder of their time being spent on tasks which were well within the capabilities of quality technicians. These results served as a baseline against which the effectiveness of efforts to improve the utilization of engineering skills could be measured.

Task Incentive Level Versus Time Allocation

The intrinsic motivational value of the job was the second factor considered. The objective of job design would be to maximize both the utilization of available skills and the level of incentive provided by the job as well. In this case the engineers were initially using a relatively high

Table 5.3 Tasks Listed in Order of Their Incentive Value

Rank Order	Task
1 (highest)	Investigate product quality problems.
2	Represent Quality Assurance in planning meetings with other organizations.
3	Review and interpret quality information.
4	Make technical recommendations on inspection methods.
5	Attend defect control meetings.
6	Determine inspection instruction requirements.
7	Review and revise process specifications.
8	Document action taken on special problems.
9	Plan analysis of inspection data for critical problems.
10	Prepare information for defect control meetings.
11	Collect and summarize quality data for special studies.
12	Prepare and verify inspection instructions.
13	Initiate corrective action to solve nonroutine problems.
14	Prepare answers to corrective action requests directed to Quality Assurance.
15	Review product planning documents.
16	Review and verify audit inspection captures.
17	Request laboratory support when required.
18	Follow up on corrective action requests directed to Quality Assurance.
19	Prepare weekly activity report.
20	Review and approve outgoing corrective action requests.
21	Prepare flow charts of manufacturing and inspection operations.
22	Provide historical data for outgoing corrective action requests.
23	Initiate corrective action requests on repetitive discrepancies.
24	Approve inspection instructions.
25	Prepare analysis for monthly product audit report.
26 (lowest)	Prepare analysis for monthly status report.

proportion of their time on high incentive tasks (the correlation between task incentive values and task times was 0.69); as a consequence, little additional gain in task incentive could be expected. The objective of job design in this case was primarily to increase the amount of time spent on higher skill level tasks. An outcome of the investigation of task incentive level was the determination of the incentive values in various kinds of quality engineering activities. Since this information may be useful in the design or redesign of quality engineering jobs, quality engineering tasks are listed in Table 5.3 in the order of their incentive value. Only the 26 most time-consuming tasks have been included in the table.

Reassignment of Tasks

The results of the task skill and time allocation measurements indicated the direction that redesign of the quality engineering job should take. Most of the technician level tasks should be assigned to quality technicians. A total of 44 quality engineering tasks had been identified by the task analysis; 22 of these had been later identified as technician tasks. To provide a means of reallocating tasks, quality technicians were obtained from a separate planning organization and were assigned to the quality engineering organization. As inspection planners they had been performing five of the 22 technician tasks. As members of the engineering organization, they retained their responsibility for the five tasks and were expected to perform as many of the other tasks as possible. The guiding assumption was that the quality technicians could obtain time for additional tasks by performing their current duties more efficiently; this was facilitated by moving them to the functional area to work directly with the engineer responsible for the area. In this way the technician had easier and more immediate access to the information and guidance that he required.

A period of approximately four months was provided for phasing the quality technicians into their new duties. During this period of familiarization and on-the-job training the assistant gradually assumed more responsibility and the amount of direction provided by the engineer was gradually reduced. An assimilation schedule was prepared for each job to ensure the systematic transfer of responsibilities.

Results of Job Design Changes

Application of the systematic job design approach resulted in significantly more effective use of engineering skills. Before the changes were made in the quality engineering job, engineers were spending only about half their time on tasks that required their expertise. After the reassignment of tasks and the phase-in of the quality technician assistants, they were devoting more than two-thirds of their time to the higher skill level engineering tasks. The gains made in more effective use of engineering talent are shown in Figure 5.2. The distribution of time alloted between engineering and technician level tasks is shown for each of the three phases of the job design effort: before job design, during technician phase-in, and after technician phase-in. It should also be noted that as a result of the changes in the jobs of the quality technicians, their skill potential was also more effectively used. Their job had been enlarged by increasing the number of tasks performed by them from five to 22.

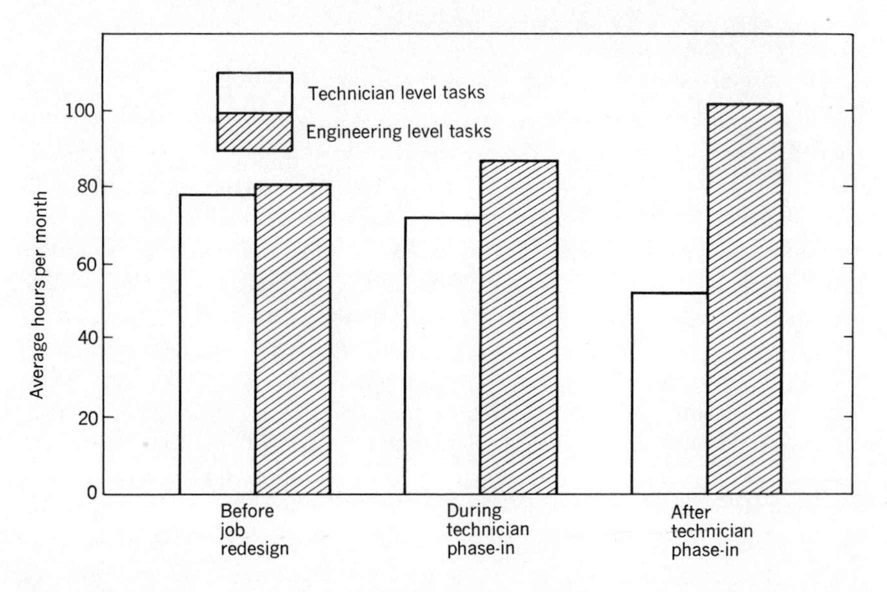

Figure 5.2 Time spent on 22 technician and 22 quality engineering tasks by quality engineers.

As expected, no significant increase was found in the time alloted to tasks having higher incentive values. The reason for this result, in this case, was the relatively high allotment of time recorded initially for those tasks with high incentive values. Maintenance of the high relationship between task incentive value and task time allocation was, therefore, about the best result that could be expected. The effect of job design changes on the relationship between task incentive value and time allocation is shown in Figure 5.3. Only the 26 tasks on which engineers spent more than one hour per month were included in this analysis. For purposes of the task incentive analysis these tasks were grouped roughly into three equal categories in terms of task incentive values.

Conclusion

As a result of developing job design techniques and applying them to a quality engineering job, it is fair to say that efforts directed toward job design can improve quality engineering effectiveness. The best use can be made of an engineer's time if he is assigned to tasks that require his specialized skills and knowledge. Care must also be taken, to include in the job those tasks that are potentially motivating to quality engineers.

ORGANIZATIONAL FACTORS IN ENGINEERING EFFECTIVENESS

The effectiveness of quality engineering is a function of organizational factors as well as the factors in engineering job design. Although the job

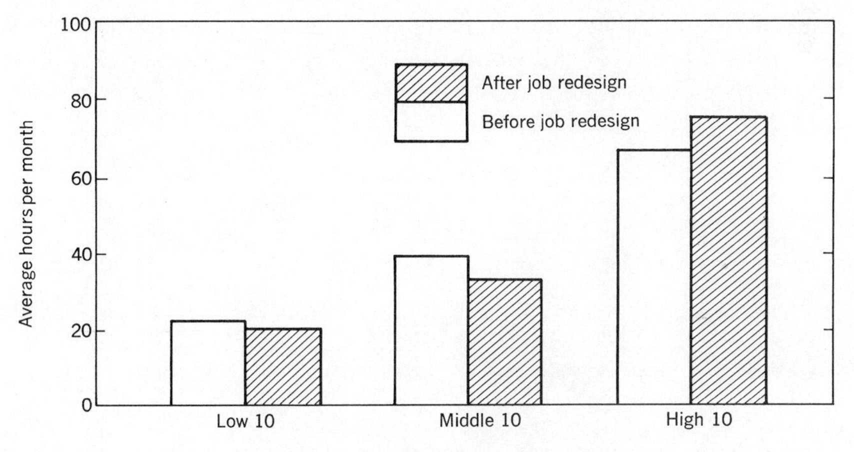

Figure 5.3 The relationship between task incentive value and time allocation before and after assignment of technician support.

is designed by allocation of objectives, functions, and tasks, the climate in which the job is performed is a function of the extent to which certain factors exist in the engineering organization. Thus a second approach to increasing the effectiveness of quality engineers is through the creation of a proper quality engineering climate. This approach complements the job design approach. In designing the engineering job, the objective was to allocate tasks in such a way that maximum utilization could be made of available skills and knowledge. When this aspect of job design has been completed, it is necessary to focus on those factors that lead to an organizational climate that enhances the performance of these tasks.

Relative Importance of Organizational Factors

An investigation of the nature and relative importance of factors which affect the performance of quality engineers was conducted in three quality engineering organizations. Initially, interviews were conducted both individually and in groups with engineers and supervisors in a variety of engineering operations. These persons were asked, generally, what aspects of their work impaired or frustrated their efforts and what aspects of their work enhanced their effectiveness and motivation. As a result of the interviews, a list of 27 factors was developed. The absence of the factor was considered to have a negative impact on engineering effectiveness; the presence of the factor was considered to have a positive impact on engineering effectiveness.

The relative importance of each factor was determined in a sample of 34 quality engineers obtained from three different departments within the Autonetics Division of the North American Rockwell Corporation. The importance of each factor was rated by each engineer with the nine-point scale shown.

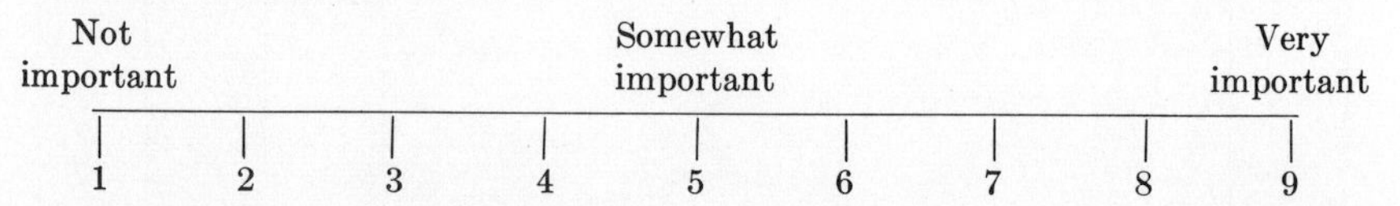

A summary of the results obtained is given in Table 5.4. The 27 factors are listed in the order of their rated importance by quality engineers and the average rating for each factor is provided in the column at the right-hand side of the table.

Improving the Quality Engineering Climate

The list of engineering effectiveness factors may be used as the basis for measuring and improving the engineering climate. By determining the extent to which each factor should exist and also the extent to which each factor does exist, an effectiveness profile for an engineering organization can be obtained. Those factors for which there are significant differences between the extent to which it should exist and the extent to which it does exist highlight areas in which the engineering climate has greatest potential for improvement. Thus the general approach that can be used for improving the climate within an engineering organization would consist of first, identification of the factors considered important in the effectiveness of the organization, second, measuring the extent to which

Table 5.4 Rated Importance of Factors Related to Quality Engineering Effectiveness (N = 34)

Rank	Factor	Rating
1	The opportunity to contribute creatively to projects to which you are assigned.	8.0
2	A feeling of worthwhile personal accomplishment.	7.9
3	Freedom in managing your own activities once an assignment has been defined and agreed upon.	7.9
4	Receiving clearly defined assignments and information about your performance from your supervisor.	7.9
5	Periodic performance reviews that provide an understanding of how you stand with your supervisor and what you should do to better your standing.	7.8
6	Knowledge that the results of your efforts will be put to a worthwhile use.	7.7
7	The opportunity to learn new things and develop new abilities in your assignments.	7.5
8	Feedback on what has become of the results of a project and how well the project objectives were met.	7.3
9	Exchanging ideas and opinions with those for whom and with whom you work.	7.3
10	The opportunity to test your abilities on increasingly difficult assignments.	7.1
11	A feeling of security and acceptance in your job.	7.0
12	A way of locating related past analyses and reports for review prior to undertaking a similar project.	7.0
13	Opportunities for professional education and development.	7.0
14	Clear understanding of the use of the end product of your efforts.	6.9
15	Sufficient orientation to understand how a particular project fits into the whole effort and how your unit fits into the engineering organization.	6.8
16	Knowing the part that your assignments play in company activities and objectives.	6.7
17	Clear understanding of how work on an assignment will be evaluated.	6.6
18	The availability of clerical and other support personnel.	6.5
19	Being able to become deeply engrossed in your assignments.	6.3
20	A relatively quiet, uncrowded space in which to work.	6.1
21	The type and amount of technical equipment and facilities.	5.9
22	The opportunity to work on several phases of a project as opposed to working on one phase of a project.	5.9
23	Time taken by the supervisor to understand you as an individual.	5.9
24	The opportunity to play a needed role in the organization (such as being the expert who is consulted on certain types of problems).	5.6
25	The opportunity to interact with management other than your own supervisor.	5.5
26	Relatively few unexpected urgent projects.	4.8
27	Freedom in selecting your own work assignments.	4.7

these factors actually exist, third, providing feedback to supervision on the existing engineering climate and areas of potential improvement, fourth, developing and initiating a plan of action for improving the climate, and fifth, remeasuring the engineering climate to determine the impact of the actions taken. This approach is illustrated in Figure 5.4.

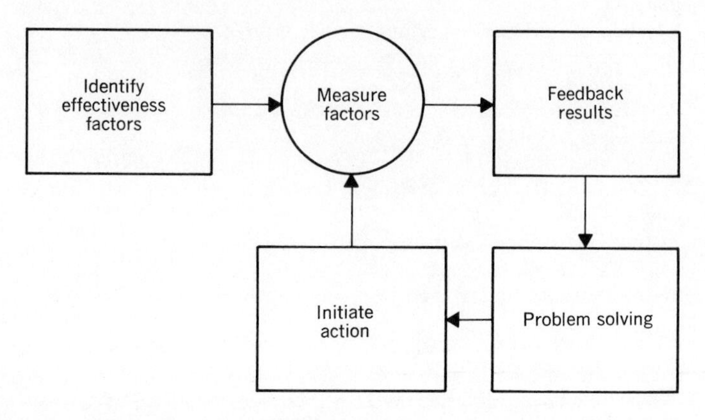

Figure 5.4 An approach to improving the climate for effective quality engineering.

A rating form such as the one shown in Figure 5.5 may be used for measuring the effectiveness climate. The form is completed by each quality engineer and technician in the organization and on the basis of the responses, average values for the extent to which each factor should exist and the extent to which it does exist are computed. In addition, using a standard statistical test, the significance of the difference between each pair of average values is computed. As a result of this analysis, any significant differences between the extent to which the factors should and do exist are identified. The significant differences indicate the factors to which the greatest attention should be given in improving the existing climate.

In developing a plan of action based upon these results, a good starting point is to discuss the results of the measures with the engineers themselves. These discussions could take place either individually or in groups. The primary idea here is to obtain the participation of the engineering group through their own efforts toward an improvement in the organizational climate. By doing so the effort will be more likely to be successful for several reasons. First, since many of the effectiveness factors included in the measure were somewhat general in nature, more detailed information is required relative to the reasons for the existence of a gap between the optimum situation and the existing situation. The best source of this information, of course, is the engineering group itself. Second and related to the first is the probability that individual engineers in the organization will have a number of good ideas for ways of improving the engineering climate. These should certainly be considered in any plan of action that is developed. Third, the involvement of the engineers themselves at the outset will assure a greater acceptance of any changes initiated as a result of the plan of action.

For each item listed below please indicate *both* the extent to which it SHOULD exist and the extent to which it DOES exist. Use the scale below as a guide. In the first box after the item put the number indicating the extent to which that factor SHOULD exist; in the second box, put the number indicating the extent to which it DOES exist.

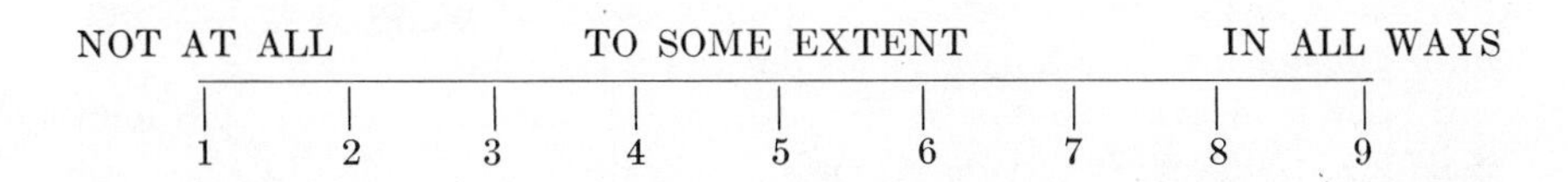

Please do not put your name on this form.

	Extent should exist	Extent does exist
1. Knowing the part that your assignments play in company activities and objectives.	☐	☐
2. The opportunity to interact with management other than your own supervisor.	☐	☐
3. The opportunity to test your abilities on increasingly difficult assignments.	☐	☐
4. The opportunity to contribute creatively to projects to which you are assigned.	☐	☐
5. Knowledge that the results of your efforts will be put to a worthwhile use.	☐	☐
6. A feeling of security and acceptance in your job.	☐	☐
7. A relatively quiet, uncrowded space in which to work.	☐	☐
8. A feeling of worthwhile personal accomplishment.	☐	☐
9. Feedback on what has become of the results of a project and how well the project objectives were met.	☐	☐
10. A way of locating related past analyses and reports for review prior to undertaking a similar project.	☐	☐
11. The opportunity to play a needed role in the organization (such as being the expert who is consulted on certain types of problems).	☐	☐

Figure 5.5 Example of part of a rating form used in measuring the climate for quality engineering.

This approach to improving the climate for quality engineering is one that would probably have its greatest impact if conducted on a continuous basis. Although it is generally recognized that the role of management is to create the kind of climate that will lead to effective performance on the part of their personnel, few attempts have been made to measure this climate in a systematic fashion. Without measures to indicate the existing organizational climate, there is little basis on which to initiate actions to improve the climate.

SUMMARY

1. The quality engineer is responsible for a wide range of activities which influence product quality and cost. These include quality planning, providing inspection information, quality monitoring, problem solving, initiating corrective action, and documenting his efforts.
2. The primary objective in structuring a quality engineering job is to maximize the amount of time spent on tasks that require engineering skill and knowledge and to minimize the amount of time spent on tasks that do not. This objective must be met in light of informational requirements and motivational considerations associated with the tasks that make up the job.
3. In estimating task-time allocation on an existing job a satisfactory technique consists of giving a task list to each of a sample of engineers and having each engineer allot 100 hours among the different tasks in accordance with how he spent his time during the previous month.
4. In an illustrative example job design techniques were applied to activities performed by a sample of quality engineers responsible for quality objectives in the production of electronic, electromechanical and mechanical devices. As a result of changes made the amount of time spent by quality engineers on tasks requiring their expertise increased from 51 percent to 67 percent.
5. The effectiveness of quality engineering is a function of factors in the engineering climate as well as factors in the design of the engineering job. A sample of quality engineers rated the relative importance of 27 factors in the engineering climate. The more important factors were directly associated with the way work assignments were handled and the quality of supervisory practices.
6. A list of organizational factors may be used as the basis for measuring and improving the engineering climate. By a determination of the differences between the extent to which each factor should and does exist areas of potential motivational improvement may be identified. The

climate for effective quality engineering can be maintained by measurements of this type followed by constructive action taken on the basis of the results.

REFERENCES

[1] G. H. Hieronymus, *Job Design: Meeting the Manpower Challenge.* Society of Personnel Administration, Pamphlet No. 15, 1965.

[2] Joseph Tiffin and E. J. McCormick, *Industrial Psychology,* 5th Ed. Englewood Cliffs, New Jersey: Prentice-Hall, 1965.

VI
THE NATURE OF INDUSTRIAL INSPECTION

Within the quality system, most of the information used for decision making and problem solving is generated by product inspection activities. As a result a significant proportion of the money allocated to quality assurance is typically spent on inspection personnel and the tools, equipment, and aids required for their work. In addition, it should be realized that inspections are performed by others than those formally identified as inspectors. Particularly in production operations of a more complex nature, production personnel must conduct inspections as an integral part of their production task. The job of an electronic assembler, for example, consists of a series of manipulations and inspections. The quality of the end product may be accounted for as much by these informal inspections as it is by the inspections conducted as part of formal quality assurance operations.

The list of inspection jobs existing in industry is likely to be considerably longer than the list of products produced. Most products are inspected in some way to assess their quality of conformance and many products are inspected for more than one aspect of quality in the course of their production cycle. Thus to discuss industrial inspection better and to provide a basis for designing inspection jobs it is necessary to identify the basic elements in inspection tasks which are common to the various kinds of inspection work that exists and is likely to exist.

BASIC ELEMENTS IN INSPECTION TASKS

There appears to be general agreement among quality assurance people and behavioral scientists about what makes up an inspection task. As a result, any inspection job can be considered to consist of a basic set of

elements. These elements will exist regardless of the particular quality characteristic inspected, the nature of the fabrication process, the nature of the quality system, or the particular tools and techniques employed by the inspector. These activity elements are interpretation, comparison, decision making, and action.

Interpretation

Every inspection job requires the interpretation of some type of established standard. A quality standard may take many forms. It may be written or unwritten; it may take the form of a general understanding, or it may be a precisely written, quantitative specification of tolerances with respect to materials, processes, methods of testing, and product characteristics. Whatever its nature, the point to be made here is that industrial inspection cannot be conducted without standards that define what is acceptable and what is not.

Comparison

Quality characteristics are compared with specified standards. If the quality characteristic is a solder connection, for example, the inspector views the connection under a microscope to determine whether or not all specified quality standards are met. The quality standards may be specified tolerances with respect to amount of solder, degree of contamination, nature of the solder flow, existence of voids, or the presence of any solder in places where it should not exist. Chances are that the comparison is made between the actual solder connection and the inspector's mental image of a defect free solder connection. Other inspections may involve much more straightforward comparisons; for example, the diameter of a hole in a machined part may be measured to determine whether it is within 0.02 inches of the specified diameter.

Decision Making

Deciding whether or not the quality characteristic conforms to the standard would appear to be a simple matter. The product either conforms or does not. There is a significant element of judgment in most inspection decisions. It may be necessary, for example, to consider the *degree* to which a quality characteristic conforms. One may wish to take a different action with respect to a solder connection that has a void but will continue to conduct electricity effectively and a solder connection with a void that is likely to cause the connection to fault under conditions of high vibration. It may thus be necessary for the inspector to indicate the degree of nonconformance as well as the existence of the nonconformance.

Action

Actions taken by the inspector on the basis of his decisions are of two basic types. The first involves some disposition of the product by the inspector. The inspector may scrap the product, reinspect it, give it to someone else for review, or perhaps do nothing with it. The second concerns recording the information obtained. As the primary source of quality data, it is necessary for the inspector to transmit his findings so that they may be useful in some manner. This information may be transmitted directly to a computer by means of a keyboard, written on an information transmittal form, or be indicated by his disposition of the product.

TYPES OF INSPECTION TASKS

The wide variety of possible inspection tasks can be classified relatively well into three basic categories. These categories are scanning tasks, measurement tasks, and monitoring tasks. Although there is always difficulty in providing a classification scheme that will handle every case, this system appears useful in discussing the design of inspection jobs.

Scanning Tasks

Probably the most common task, at least in past years, and the one most people seem to think of with respect to inspection operations is the scanning task. In this type of inspection the inspector searches for defects by making a point by point examination of an item. Usually, his examination is performed visually, although other senses may also be employed from time to time; for example, a scanning inspection task may consist of examining the quality of a machined part visually for most imperfections but rubbing the finger over some surfaces to identify nicks and burrs. Scanning tasks are included in a wide variety of inspection jobs. Included are the examination of electronic hardware for improper connections, wiring, component placement, and contamination; the inspection of fruit, vegetables, peanuts, and so on for the purpose of culling out items or classifying by grades; the inspection of automobile body surfaces for blemishes in paint; and the examination, by lot, of any number of machine-produced items for purposes of identifying and enumerating defective items.

Measurement Tasks

The measurement task category includes inspections in which the inspector measures the dimensions of products using a measuring instrument to determine whether or not the measured dimension is within specified tolerances. The complexity of a measurement task may vary greatly. The task may consist of something as simple as the application of a go, no-go gauge. On the other hand, it may involve an extensive calculation in preparation for taking the measurement and the use of a complex measuring instrument. A relatively common task of this sort is in the inspection of parts that must be machined to precise tolerances. The skills and knowledge required for conducting machined part inspections may require many years of training and experience to develop the required proficiency.

Monitoring Tasks

Monitoring tasks are typically associated with the control of some type of automatic or semiautomatic equipment. The task of the inspector is to observe displays for indications of out-of-tolerance conditions. One basic difference between monitoring tasks and measurement or scanning tasks is that the inspector typically does not deal with the product directly; for example, the inspector or test operator observes dials, meters, and printed readouts during the test of an item of electronic equipment but does not observe the item of equipment directly. In another example, although he is not called an inspector, a member of the flight crew of an intercontinental jet aircraft observes the condition of his engines by monitoring a panel of indicators. His indication of a defective or unsatisfactory condition is an abnormal reading by one or more of his indicators. With the trend toward increased automation, it is likely that this type of inspection task will become increasingly common. In fact, as the automation of industrial processes increase, it is likely that many people who now perform manipulative type operations in industry will be replaced in large numbers by people who perform inspection operations of this type.

FACTORS IN INSPECTION ACCURACY

The organization of inspection activities and the development of procedures and tools needed to assure accurate inspection performance require an understanding of the factors that influence inspection accuracy. There are three basic types of factors involved: those associated with the individual abilities of the inspectors, those associated with the physical nature of the inspection task and the surroundings in which the inspection is conducted, and those relating to the organization and methods that define the inspection job. The last category includes the interpersonal relations of inspectors with others including other inspectors, supervisors, and manufacturing personnel. Using a slightly different classification system, McKenzie [1] emphasized the importance of interpersonal relations on inspection accuracy. He provided evidence to support the idea that inspectors are frequently influenced to a greater extent in their inspections by other people than they are by the quality level of the product inspected. Consequently it is somewhat difficult to isolate the effect of any given factor because there is typically an interaction among factors and it is this combined effect that we typically observe when investigating inspection accuracy. In spite of this basic difficulty, however, in investigating inspection accuracy, several controlled studies have been conducted which provide information which is useful in the design of inspection jobs.

The Effect of Complexity

The complexity of the inspection task has a significant effect on inspection performance. This was illustrated by a study of the effect of the complexity of electronic equipment on inspection performance [2]. As an attribute of electronic equipment, complexity was considered a function of the number of parts that comprised an equipment item and the way the parts were interrelated or packaged. Complexity was considered to increase both as the number of parts increased and as the arrangement of parts became less orderly.

It was thought that inspectors would be less effective in detecting defects within more complex items than in less complex items; that is, the more complex the item, the smaller the percentage of defects detected. Moreover, that when more parts of different types are involved, more quality characteristics are involved. Therefore the inspector must be aware of more standards upon which to base his judgments of conformance. Also, confusion may be more likely to occur when he is searching a more complex item for defects. In short, the more complex the item, the more complex the inspection task.

The objective of the study was to determine the relationship between complexity and inspection accuracy. The approach taken was to measure the accuracy of inspections of each of ten different items of electronic equipment, measure the complexity of each item, and determine the relationship between the accuracy and complexity measures. Ten items of electronic equipment were inspected by 62 experienced inspectors. Eight or more inspectors inspected each item; each inspector had an unlimited amount of time to make his inspections.

The measure of inspection performance associated with each equipment item was the percentage of defects detected in the item by the eight or more inspectors who inspected it. Defects in each item had been

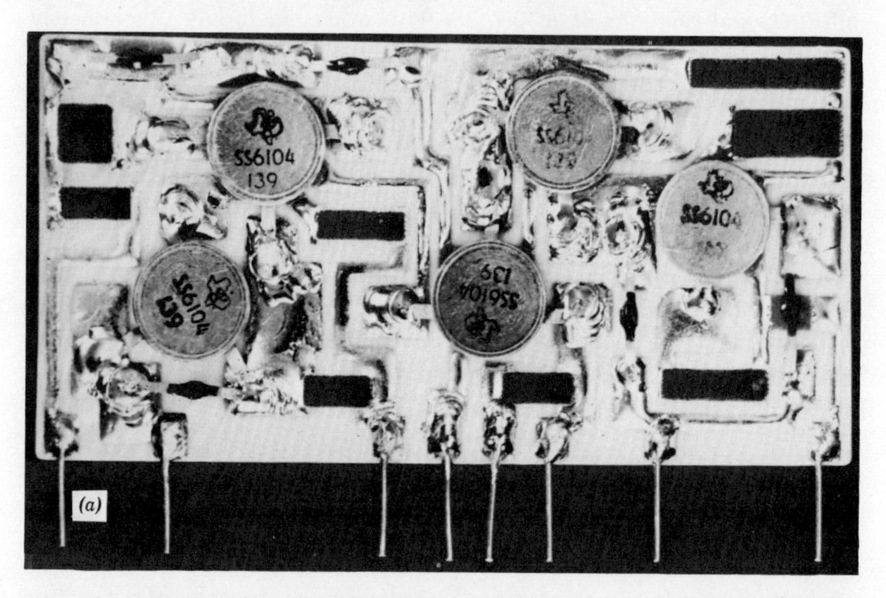

Figure 6.1 Sample electronic items and their complexity indexes: (*a*) ceramic printed circuit, complexity index = 28; (*b*) autonavigation module, complexity index = 84.

verified by a panel of four or more experts. Equipment complexity was measured by counting the number of major parts comprising each item, for example, circuit boards, resistors, wire bundles, connectors, and transistors. The complexity index thus obtained ranged from six to 100. Two of the items together with their complexity indices are presented in Figure 6.1.

The results indicated that complexity has a significant negative effect on inspection performance which apparently cannot be overcome simply by providing inspectors with an unlimited amount of inspection time. This finding suggests that inspection jobs must be designed so as to overcome the effect of task complexity; significant gains in inspection performance may be obtained by reducing task complexity or by developing procedures and aids which reduce the effect of complexity. The relationship between inspection performance and equipment complexity is shown in Figure 6.2.

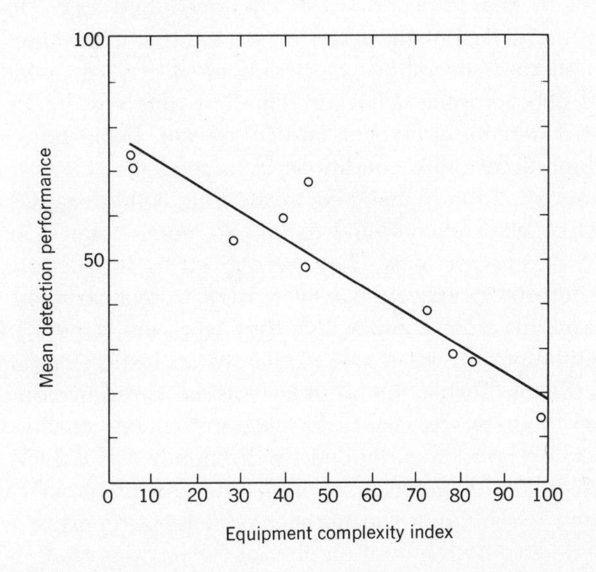

Figure 6.2 Inspection performance as a function equipment complexity.

The Effect of Defect Rate

Most production programs have a typical learning curve associated with them. Early in the program, product defect rate is high but quality usually improves until, later on, the defect rate is relatively low. Other circumstances, such as variations in production schedules, may also cause the defect rate to vary in a predictable way. Should inspections be conducted in the same manner regardless of the defect rate? This depends, of course, on the relationship between defect rate and inspection accuracy. A study was conducted to determine the effect of defect rate on inspection accuracy [3].

Since both scanning and monitoring inspection tasks have some elements in common with the tasks studied in vigilance research, which required the detection of infrequent signals appearing at irregular intervals, previous research on vigilance performance may suggest the general nature of this relationship. One would predict from vigilance research that inspection accuracy would decrease with a decrease in defect rate [4]. However, the differences between most industrial inspection tasks and the tasks used in the referenced research suggested the need for further investigation.

A representative scanning type inspection task was developed and materials prepared which included defects at four different defect rates. Performance on the inspection task employed had been found to be significantly correlated with performance on the inspection of inertial instruments, module assemblies, electronic circuit boards, microelectronic devices, and photographic materials. The four different defect rates were 0.25 percent, 1 percent, 4 percent, and 16 percent. Inspections were made under the four defect rate conditions by a total of 80 naïve inspectors. Inspectors were randomly assigned to the four conditions, 20 per condition. Inspectors were encouraged to take as much time as necessary to perform accurate inspections. They were given no instructions about the number of defects to expect; however, they were instructed to ask for clarification on any item about which they were undecided. In this manner the monitor performed a role similar to an inspection supervisor or leadman on the job. Inspection accuracy was measured in terms of defects detected and false reports made. For each inspection condition the number of defects detected was divided by the number of defects present to give the percentage of defects detected. The percentage of false reports was computed by dividing the number of defects reported which were not actually defects by the total number of defects reported.

Inspection accuracy decreased with reductions in defect rate. As shown in Figure 6.3, the percentage of defects detected decreased slightly between defect rates of 16 percent and 1 percent, but dropped sharply between 1 percent and 0.25 percent. False reports, the second indicator of inspection accuracy, became more frequent with reductions in defect rate. The percentage of false reports increased at an accelerated rate as the defect rate approached zero. The differences among the four defect rates for the percentages of defects detected and for the percentages of false reports were found to be statistically significant beyond the 0.05 level for both cases.

These results demonstrate a potential problem in using information obtained from industrial inspections and have implications for management of inspection operations. When a relatively high quality product is being produced (the percentage of defective characteristics is about one

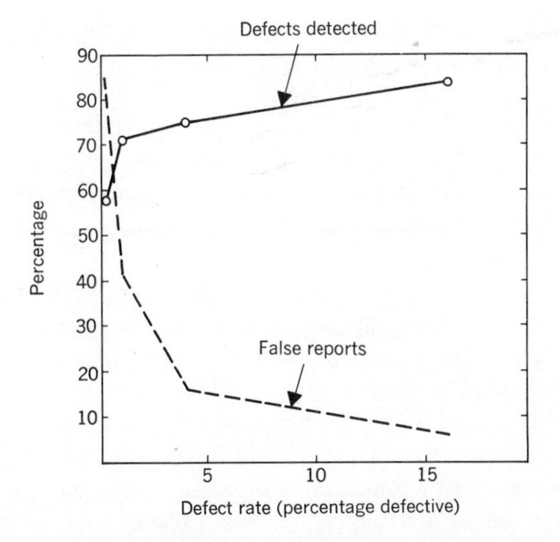

Figure 6.3 The relationship between defect rate and inspection accuracy.

or less), inspections may not be very accurate. The percentage of defects detected is likely to be low and the percentage of false reports is likely to be high. Consequently, inferences made about the true quality level may be in error. Thus supervisors should be aware that this tendency exists and be prepared to counteract the effect of this factor; for example, as quality level increases, considerations might be given to changing the nature of the inspection job. Fewer and more intensive inspections might be made through the use of sampling plans. Much better information might be obtained at less cost, for example, from inspections of a 10 percent random sample of the product than from inspections of 100 percent of the product.

The Effect of Repeated Inspections

A solution sometimes offered to problems of low inspection accuracy or of obtaining extra high accuracy in sensitive quality areas is the initiation of product screening by manufacturing personnel or repeated inspections by quality assurance personnel. The assumption is that the likelihood of detecting defects increases with the number of different inspections made. How effective is this approach? When the same item is inspected by different inspectors, it is possible that each inspector may be detecting the same more easily detectable defects. On the other hand, independent inspections of the same item by different inspectors may add significantly to the overall level of inspection accuracy. A study was conducted at the Autonetics Division of the North American Rockwell Corporation to obtain some initial information about the effect of repeated inspections on inspection accuracy.

Two electronic module assemblies which had been prepared for final inspection were used in the study. The defects present in each assembly had been determined from independent inspections by a panel of quality assurance experts. In addition, several defects were added so that each assembly contained defects representative of the types of defects typically encountered by inspectors.

Ten experienced inspectors each independently inspected the two module assemblies. These inspectors made up the group regularly assigned to the inspection of module assemblies of this type. To investigate the effect of repeated inspections on inspection accuracy, the percentage of master list defects detected was computed for each possible number of repeated inspections—1, 2, 3, 4, 5, 6, 7, 8, 9, and 10.

Repeated inspections significantly increased inspection accuracy up to a point. Performance in detecting critical defects continued to improve (at a slightly decreasing rate) with the addition of independent inspections up to a total of six. Little increase in inspection accuracy was noted when more than six independent inspections were employed. Performance in detecting noncritical defects, on the other hand, continued to improve as additional inspections were added. The overall performance level in detecting noncritical defects was lower and the rate at which accuracy increased was also lower than for the detection of critical defects. These results are shown in Figure 6.4.

The results of this study suggest that repeated inspections might be considered as a means of increasing inspection accuracy under certain conditions. These results should be regarded as preliminary, however, because this study included only one type of product and one set of inspectors. The relationship between inspection accuracy and the number of repeated inspections may be different in different situations. Yet these results clearly show that different inspectors do not necessarily find the same defects and, as a consequence, a team approach to inspection may be useful under certain conditions. They also suggest that the employment of screening personnel may be of value in certain manufacturing operations.

Vigilance Factors

Although it can be said that vigilance is necessary in the successful performance of each of the three types of inspection tasks, most of the research on vigilance performance has centered around situations in which human beings are required to monitor some display in search of critical but infrequent signals. Therefore most of the tasks studied in vigilance research must be classified as monitoring tasks. Thus the information ob-

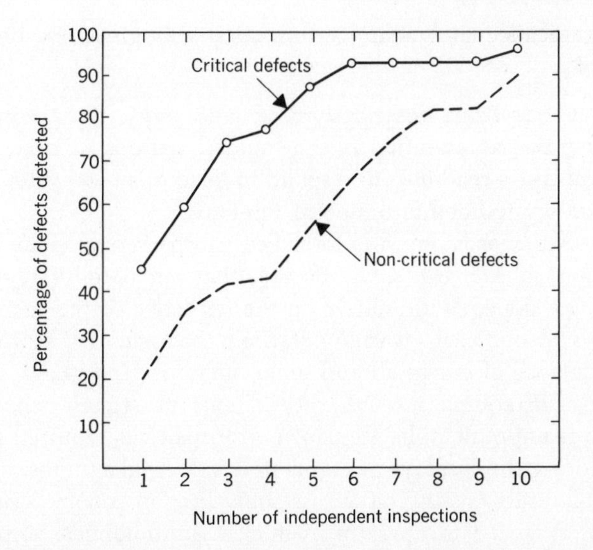

Figure 6.4 Inspection accuracy as a function of number of independent inspections.

tained from vigilance research is most directly relevant to monitoring type inspection tasks. The application of these findings to the other two types of inspection task, measurement and scanning, should be done with a considerable amount of caution.

An additional reason for caution in applying the body of knowledge which has been obtained from vigilance research to the design of industrial inspection jobs is that much of this research has been conducted in the laboratory. In a paper that discussed the consideration of industrial inspection as a vigilance task Baker [5] indicated that the application and evaluation of laboratory findings was needed before they could be applied directly to the design of industrial inspection jobs. Even with the degree of caution required, however, he concluded that by knowing about behavior in laboratory vigilance tasks a significant amount is also known about the behavior of inspectors of industrial products.

The term vigilance has somewhat different meanings to different researchers, depending on the context in which it has been studied, the nature of the tasks employed in research studies, the research methods that have been employed, and the theories that have been generated to explain the research results. In spite of this, however, a significant amount of information relevant to industrial inspection has been developed. Some of this information may be useful in the design of industrial inspection jobs.

One of the clearest recent statements on the nature of vigilance and its significance in human performance was given by Buckner and McGrath [6].

"Vigilance has been described in several ways: As performance on monitoring tasks, as attention over extended periods of time, as a state of the organism—a readiness to respond to infrequent low intensity signals occurring at unpredictable temporal intervals.

"Vigilant behavior is required for successful performance in a variety of situations and jobs. It can safely be said that our defense against all out nuclear attack depends ultimately on the vigilance of the men observing the displays in our early warning stations throughout the world. It depends on the skill of our radar and sonar operators and image interpreters in detecting infrequent low-intensity signals or stimuli whose time of occurrence is unpredictable. Vigilant performance is required in industry —in monitoring the displays of automated industrial systems, in detecting unacceptable units coming off an assembly line, in proofreading a manuscript. It is required in operating vehicles—automobiles, ships, aircraft, even spacecraft. To perform successfully operators must be alert for indications of malfunctions, changes in operating states and relevant stimuli in the operating environment.

"The increased sophistication of electromechanical systems in modern society tends to reduce man's role as an operator and to increase his role as a monitor. It tends to emphasize the importance of vigilant behavior and the importance of our understanding more about it."

In spite of the significance of vigilance to industrial inspection, the application of findings from research on vigilance to the design of inspection jobs should be approached with considerable caution. Most vigilance research has been conducted in the laboratory with simplified forms of monitoring type inspection tasks. Typically, the defect or signal to be detected was easily recognizable and required little or nothing in terms of a standard for comparison and thus little judgment on the part of the inspector. As a consequence these vigilance tasks usually have not included all of the elements characteristic of industrial inspection tasks. The kind of difference these omissions can make was illustrated in a study by Jerison [7].

A relatively consistent finding in vigilance research is that the percentage of signals or defects detected decreases with the passage of time at the task. A relatively high percentage of signals are detected initially but this level of performance has been found to drop significantly during the first 15 to 30 minutes on the job. Occasionally, however, a study is conducted in which this decrement in performance is not found. Jerison's study investigated one factor that seemed to have an effect on this vigilance decrement phenomenon. This factor was the complexity of the task.

In his study he used a typical vigilance task consisting of a simple clocklike apparatus in which a hand moved ahead typically in single jumps. It was programmed, however, to move ahead with a double jump at infrequent and irregular times. The task of the inspector was to detect the double jump of the clock hand. For purposes of the study the complexity of this task was increased by the use of additional clocks; detection performance was measured under conditions in which one clock, two clocks, and three clocks were used. Under each condition inspectors monitored the clock display and reported any double jumps in the clock hand that they observed during a two-hour period. With the more typical vigilance task (the one-clock condition), the typical decrement in performance was observed. However, no decrease in the percentage of signals detected was observed under either of the two more complex task conditions. As shown in Figure 6.5, although the overall performance level was lower for the two-clock condition and lower yet for the three-clock condition, performance measured at the start of the two-hour period was no better nor worse than performance measured at the end of the two-hour period.

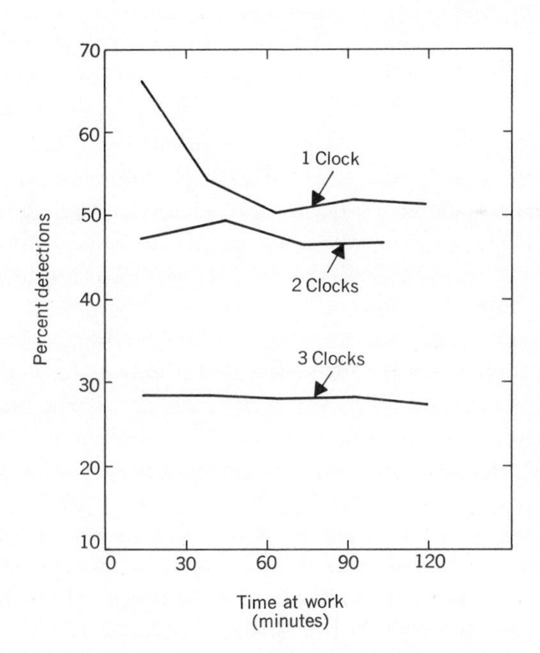

Figure 6.5 Percentage of detections during successive time periods with one, two, or three clocks in the monitored display (adapted from Harry J. Jerison, On the Decrement Function in Human Vigilance. In D. N. Buckner and J. J. McGrath (Eds.), *Vigilance:* A Symposium. New York: McGraw-Hill, 1963. Pp. 199–216.)

The implication is that as the laboratory vigilance task more nearly approached a monitoring type inspection task, findings from vigilance research became less applicable. This is not to say that findings from vigilance research are never applicable to inspection tasks. The point is that most inspection tasks are typically more complex and more intellectually demanding than the tasks studied in vigilance research and, as a result, the findings from vigilance research should be applied to industrial inspection tasks only with considerable care.

SUMMARY

1. The basic elements in an inspection task are interpretation of the specified quality standard, comparison of the quality characteristic with the specified standard, deciding whether the quality characteristic conforms to the standard, and action taken as a result of the findings.

2. There are three basic types of inspection task: scanning, measurement, and monitoring. In a scanning task the inspector searches for defects by making a point by point examination. Measurement tasks include inspections in which the inspector, using a measuring instrument, measures the dimensions of products to determine whether or not the measured dimension is within specified tolerances. Monitoring tasks are typically associated with the control of some type of automatic or semiautomatic equipment or system. The task of the inspector is to observe displays for indications and initiate action using appropriate controls.

3. There are three basic types of factor that influence inspection accuracy: those associated with the individual abilities of the inspector, those associated with the physical nature of the inspection task and the work environment, and those that relate to the organization and methods that define the inspection job.

4. The complexity of the inspection task has been found to have a significant negative effect on inspection performance. This effect is apparently one that cannot be overcome simply by providing inspectors with an unlimited amount of inspection time.

5. Inspection accuracy has been found to decrease with reductions in defect rate. Thus, when a relatively high quality product is being produced (the percentage of defective characteristics is about one or less), inspections may not be very accurate. The percentage of defects detected is likely to be low and the percentage of false reports is likely to be high.

6. Repeated inspections may be an effective means of increasing inspection accuracy under certain conditions. Initial results indicated that different inspectors do not necessarily find the same defects and, as a result, a team approach to inspection may be a solution to problems of

low inspection accuracy or of obtaining extra high accuracy in sensitive quality areas.

7. Since many of the tasks studied in vigilance research do not contain all elements of inspection tasks, caution should be exercised in the application of results from vigilance studies to the design of inspection jobs.

REFERENCES

[1] R. M. McKenzie, On the Accuracy of Inspectors. *Ergonomics,* 1958, **1,** 258–272.

[2] Douglas H. Harris, The Effect of Equipment Complexity on Inspection Performance. *J. appl. Psychol.*, 1966, **50**, 236–237.

[3] Douglas H. Harris, Effect of Defect Rate on Inspection Accuracy. *J. appl. Psychol.* (in press).

[4] James J. McGrath and Albert Harabedian, Signal Detection as a Function of Inter-Signal-Interval Duration. In D. N. Buckner and J. J. McGrath (Eds.), *Vigilance: A Symposium.* New York: McGraw-Hill, 1963. Pp. 102–113.

[5] C. H. Baker, Industrial Inspection Considered as a Vigilance Task. Paper presented at the Fifteenth International Congress of Applied Psychology, Ljubljana, Yugoslavia, 1964.

[6] D. N. Buckner and J. J. McGrath (Eds.), *Vigilance: A Symposium.* New York: McGraw-Hill, 1963.

[7] Harry J. Jerison, On the Decrement Function in Human Vigilance. In D. N. Buckner and J. J. McGrath (Eds.), *Vigilance: A Symposium.* New York: McGraw-Hill, 1963. Pp. 199–216.

VII

DESIGNING THE INSPECTION JOB

Considerations in designing inspection jobs are discussed in this section. With the description of the nature of inspection work as background, specific aspects of inspection job design are presented. These include the organization of inspection work, preparation of inspection instructions, design of the workspace, and the design of displays and controls for monitoring tasks. Also, some specific information and techniques that may be used in designing inspection jobs are presented.

ORGANIZATION OF INSPECTION WORK

Inspection work should be organized to maximize the effectiveness of the quality system in meeting quality objectives. In so doing several aspects of inspection work are to be considered. They include the way quality characteristics are assigned to inspection stations, the quality standards which apply to each characteristic at each station, the particular tools and techniques made available at each inspection station, and the methods of communication used among inspectors, inspection supervisors, manufacturing personnel, and others involved in the production process.

Inspection Stations

In many respects the inspection station defines the job of the inspector. An inspection job consists of a group of tasks that, in turn, are determined by the particular set of quality characteristics to be inspected, the standards to which they are to be inspected, when and how the inspections are conducted, the tools and techniques employed, and the communications that take place in the process. Certain of these aspects are dictated by the particular location or placement of the inspection station in the production process; for example, the number and location of inspection stations determine the number of quality characteristics to be considered by each inspector. It was shown by the study of the effect of complexity on inspection effectiveness that as more characteristics are considered, inspections become less effective. In organizing inspection work an attempt should be made to minimize the number of quality characteristics considered at each inspection station within the limitations of costs and available personnel. Furthermore, periodic reviews should be made of inspection stations to determine the number of quality characteristics considered at each station and an attempt made to organize inspection work to minimize the number of characteristics for which any one inspector is responsible.

The physical location of inspection stations may also play an important part in the organization of inspection work. Inspection tasks are demanding in terms of the effort and concentration they require. As can be seen in Figure 6.2, the average percentage of defects detected in the more complex items was less than 50 percent for this sample of trained, experienced inspectors. These measures were obtained under conditions in which the inspectors were not disturbed during their inspections and in which they had as much time as they required to complete their inspections. Measures obtained subsequently under normal inspection conditions for certain of these same items indicated that inspections made on the job were significantly less accurate than the inspections made under the more optimal conditions. It would seem appropriate to locate inspection stations where the amount of disturbing influence is minimized and to minimize unrealistic time pressures which might result from locating inspection stations too close to the mainstream of production activity.

Quality Standards

Also to be considered in the organization of inspection work are the quality standards used by each inspector. The quality standards and the quality characteristics to be inspected are, of course, closely related. Consistent with the objective of minimizing inspection task complexity is the attempt to keep the number of different quality standards used by any given inspector to a minimum. The best situation, but one that is not usually feasible, would be to organize inspection work so that an inspector inspects to only one standard at any point in time. When using a number of quality standards at a time, there is too much opportunity for negative transfer of training effects to occur. This was illustrated by the results of a study of the use of quality standards in several companies [1]. Inspecting to more than one set of standards was found to adversely affect inspection performance. In one company the same inspectors were used on both commercial and government work. Performance on both deteriorated even though distinct specifications and procedures were provided for each kind of work. People were found to work best when they were required to work to one set of standards on a continuing basis.

Tools and Techniques

Inspection performance may be enhanced by organizing inspection work in a manner that will permit the use of the most effective tools and techniques. During the initial design of inspection jobs a review should be made of developments in inspection methods. Innovations which appear to be appropriate to the job should be evaluated and those found to be of value incorporated into the design of the job. Since the tools and techniques employed will have an impact on the manner in which personnel are assigned, on the definition of quality standards and on the manner in which inspectors are instructed, tools cannot be considered independently of these other aspects of the inspection job. A detailed discussion of inspection tools and techniques together with human factors research and development efforts leading to them, are presented in Chapter IX.

Communications

An important principle to apply in organizing inspection work is to give the inspector only the information he needs and to obtain from the inspector only the information he can give. The difficulty in attempting to apply such a principle is determining how much information an inspector needs and how much information should be obtained from him. The intent here is not one of establishing guidelines for information requirements because they will vary in different situations. However, some new dimensions in communications have been shown in recent human factors research to have a significant impact on the organization of inspection work. They are discussed briefly here and in detail in Chapter XIII.

The only communications normally considered in the inspection job are those associated with providing information such as quality standards and procedural instructions to the inspector and obtaining quality data from the inspector. Normally not considered are those communications that are associated with inspector motivation and that consequently may have a significant impact on how well the inspection work is performed. They are outlined here because their effectiveness depends on how well communication channels are established in the organization of inspections work. Among communication channels, three primary ones are considered: a channel to permit inspectors to suggest ways of reorganizing inspection work, a channel used to establish inspection goals, and a channel in which inspectors may learn how well they are performing their inspection tasks.

The inspector himself may be a rich source of ideas about how his job might be improved. Chances are, however, that he will not even think of job improvements unless he is aware of some means of bringing his ideas before someone who can judge their worth and have some impact on their implementation. A very effective means is an indication by the supervisor to his inspectors that he is receptive to ideas for job improvement and is continually reinforcing the contributors by positive action. Unfortunately, the supervisor often becomes engrossed in the day-to-day involvements of his job; as a result, this communication channel is frequently nonexistent. For this reason it is sometimes well to consider a more formal channel, one that would require the inspector to jot his idea down on a form and submit it to an established board of review. Such a system should be characterized by rapid feedback to the individual, noting the disposition of his idea. It also should be established to preclude a feeling of circumventing the supervisor. The main value of the more formal system is its continued existence and thus the continued reminder it provides that ideas for improvement are desired.

A second motivational communication channel is one that will provide a translation of company goals into the specific goals of the inspection group. Some specific techniques in communication of this sort are discussed at greater length in Chapter XIII together with the findings of recent research. The basic idea behind goal setting is that nearly all human behavior is goal directed in one fashion or another; therefore it would be quite unnatural for a group of inspectors to be working without some specific and realistic goals in mind. In organizing inspection work attention should be given to goals acceptable to the inspection work group and the method by which they are established.

A third aspect of communications is providing each individual inspector with information on how well he is performing his job. The motivation aspect of this form of communication is also discussed in Chapter XIII. This communication channel, a feedback loop in system terminology, should be considered as much a part of the inspection system as the tools and procedures the inspector uses. Without the feedback loop to tell the inspector specifically the kinds of error he has a tendency to make and to tell him how far he is from his inspection goals, we have no reason to expect improvement in his performance.

INSPECTION INSTRUCTIONS

Effective inspection performance depends on both the adequacy of information provided to the inspector and the degree to which the in-

spector understands and uses the information. Adequate information that is not used has precisely the same effect as no information at all. Consequently, it is as important to provide inspection information in a form that will be useful as it is to make sure that the information is complete and accurate. The written instruction is one of several possible forms that may be used in providing inspection information. The types of information required by the inspector and several principles which, when applied, should result in usable written instructions are discussed in the following paragraphs.

Types of Information Needed

All written inspection instructions should include one or more of the following six basic types of information needed by the inspector:

1. *When to inspect.* The inspector needs to know what operations come before and after his inspection.
2. *What to inspect.* The inspector needs to know what he is to look for during his inspection.
3. *Inspection standards.* The inspector needs to know what is acceptable and what is unacceptable for each characteristic inspected.
4. *Inspection methods.* The inspector needs to know the procedures that will result in most effective inspections and that will permit reliable interpretations of the results of his inspections.
5. *How to report data.* The inspector needs to know the data recording and reporting procedures so that the information he provides can be reliably interpreted.
6. *Material handling.* The inspector needs to know the appropriate actions to take with the materials he handles.
7. *Administrative procedures.* The inspector needs to know the rules and procedures which govern his conduct on the job.

Principles for Writing Inspection Instructions

The effectiveness of written instructions in obtaining the desired inspection performance depends on the degree to which they are complete and understood. The following principles are included as guidelines for writing inspection instructions. The examples were taken from existing instructions.

1. *A specific objective should be established for each instruction.* Unless the writer determines the inspection action desired, the instructions he writes probably will not result in that action. The objective may be stated in terms of one or more of the six basic information requirements discussed earlier; for example, an instruction telling an inspector what to inspect might be written, "Inspect mechanical wrap." In terms of the desired objective, the instruction, "Verify that there is no evidence of poor or incorrect mechanical wrap," contains no useful information not contained in the shorter instruction.

2. *The correct inspection method should be stated in operational terms.* The following example of an operational instruction tells the inspector exactly what to do.

"Center the part on the meter table with its longest dimension parallel to the table. While rotating the part about its normal axis, record the peak-to-peak flux meter reading."

On the other hand, "Determine the flux meter reading of the part per specification," does not tell the inspector what operations to perform.

3. *Only necessary words should be used.* Only words that are necessary to obtain the desired result should be included in the instruction. Besides wasting the inspector's time, unnecessary words may tend to confuse the inspector or hide the significant information; for example, consider the following instruction:

"Parts shall be inspected microscopically to insure cleanliness. Particular attention will be directed to the longitudinal and radial holes in the shaft, also thrust surfaces will be inspected for nicks, scratches and burrs. These areas lie in, or constitute the fluid bearing flotation system and must be immaculately clean,"

This conveys less information than the following:

"Using 15X magnification, inspect shaft holes for cleanliness and thrust surfaces for nicks, scratches and burrs. CAUTION Cleanliness is critical in these areas."

4. *Adjectives and adverbs should not be used.* Modifying words such as good, poor, incorrect, clearly, utmost, carefully, gradually, thorough, complete, and securely add nothing to written instructions because they are not precise. Such words as excessive and properly will have different meanings to different inspectors. Adjectives and adverbs must be replaced by statements more specific or quantitative in nature. For example, "Inspect to assure that the nameplate has been correctly installed," should be replaced with, "Inspect nameplate. Location should be 3 inches from the top and 4 inches from the left side of cover. Nameplate should resist prying up with fingernail."

5. *Only common words should be used.* When selecting alternative words that have about the same meaning, select the word with the widest use. In so doing the most understandable word should be selected. Fancy words also may be avoided by using several smaller words instead. Some "twenty-dollar" words found in existing inspection instructions and the "ten-cent" alternative to each are listed:

Avoid	*Use*
parameters	values
warrant	call for
scrutinize	look at
incorporate	put
delineate	tell
verification	check
precede	come before
subsequent	later
facilitate	make easy
indicative	points out

WORKSPACE DESIGN

In designing the inspection workspace, both the activities to be performed and the characteristics of the inspector need to be considered. The recommended approach to laying out the workspace includes obtaining detailed information on the tasks to be performed and determining the general nature and environment of the workspace in light of the physical characteristics of the inspectors who will be using it. In addition to providing the specific tasks to be performed by the inspector, the task analysis will provide the sequence of operations, the location of any equipment and tools required, the location of inspection instructions, and environmental considerations such as illumination and the operator's body position during the inspection operations. After these types of information have been obtained, the general layout of the inspection workspace becomes a relatively simple matter. Unfortunately the inspection workspace is typically designed with little attention given to this important first step. Frequently overlooked, for example, is provision for adequate display of inspection instructions. There is often neither sufficient space provided for the instructions nor an effective means of integrating the instructions with the inspection tasks for which they are provided. The usual result of this oversight is increased inspection time and errors due to difficulty experienced in using the instructions or due to not using the instructions at all.

Workspace Dimensions

There are several excellent human factors engineering handbooks [2,3,4] that can be consulted in determining the workspace dimensions. These handbooks discuss factors to be considered in workspace design and provide workspace dimensions appropriate for specified work conditions and segments of the worker population. Typically, workspace dimensions are provided to accommodate both the fifth percentile (small operator) and the 95th percentile (large operator) individuals. In this manner workspace dimensions that are established will accommodate most of the operator population. Of course, there may be several different possible operator populations. For example, a workspace for an inspection job that will be performed only by women would differ in dimensions from a workspace that would have to accommodate any citizen of the United States. The more specifically the operator population can be described, the more optimally the workspace can be designed for the people who will actually be using it.

The three primary body positions to be considered in designing the

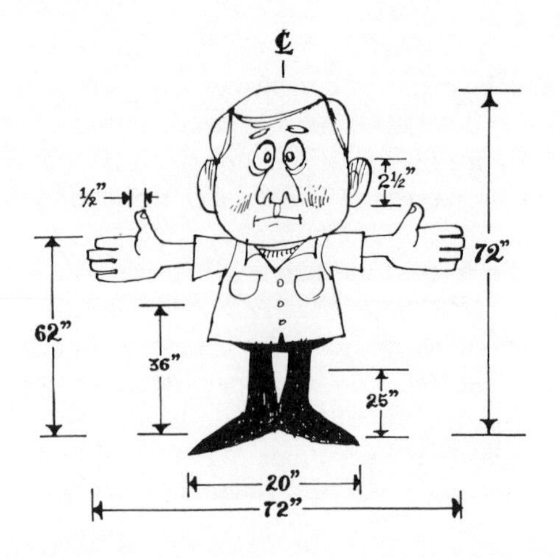

workspace for inspectors are associated with seated operations, standing operations, or both seated and standing operations. Workspace dimensions and configurations will differ among these three cases. As a quick check on the acceptability of workspace provisions for men and women engaged in each of these three types of operation, the following list is provided.

For seated operations the workspace dimensions are

bench height	30 inches
seat height	17 inches, adjustable ±2 inches
knee and foot room	25 inches high
	20 inches wide
	18 inches deep

For tasks which include writing or recording data, the minimum size bench surface is 16 inches deep and 30 inches wide. Additional space is required for tools, manuals, worksheets, aids, and equipment.

For standing operations the workspace is 36 inches from the floor; a kick space measuring a minimum of 4 inches deep is required.

For operations which require both sitting and standing activities, the workspace dimensions are

seat height	34 inches, adjustable ±3 inches
foot rest	16 inches, adjustable ±4 inches

Workspace Illumination

The primary visual considerations in designing an inspection workspace is the illumination and how it is provided. An extensive amount of research on illumination requirements for various tasks has been conducted. For example, the Illuminating Engineering Society has sponsored research on a number of aspects of workspace illumination. The results of these studies have been reported in the Society's journal, *Illuminating Engineering*. A particularly notable research effort was conducted to develop techniques for specifying interior illumination levels for visual tasks [5,6]. Since the findings and recommendations from these and similar studies are provided in detail elsewhere, the attempt here is to briefly summarize findings which are relevant to the design of inspection workspaces. The two aspects of illumination considered are illumination level and light distribution.

In general, the level of illumination desired for an inspection task is that which will provide the greatest inspection accuracy. The precise determination of illumination level in terms of performance measures is difficult to do, however, short of conducting an extensive experimental study for each task. As a consequence attempts have been made to study a range of different types of visual tasks to establish a set of bench marks. This approach provides a means for roughly establishing the illumination level for a task in terms of certain specified task characteristics. A crude set of bench marks for visual inspection tasks was developed from a review of specified illumination requirements designed for a variety of visual task categories. These recommended illumination levels are provided in Table 7.1.

Table 7.1 Recommended Illumination Levels for Inspection Tasks

Type of Work	Foot-Candles
Unmagnified visual, functional, and dimensional product inspection.	100
Large area magnification for inspection of small details frequently requiring low power magnification.	200
Microscopic examination of materials, surfaces, and finishes usually requiring spot illumination.	500
Highly magnified examination of materials and small details always requiring high intensity special lighting.	1000

The primary consideration, with respect to the manner in which light is distributed, is the trade-off between the illumination level required and any glare that may be produced. Glare is brightness within the visual field which produces discomfort or interferes with vision. It is easy to see that the possibility of glare is greatest with the higher levels of illumination. Thus the problem is one of how to maximize illumination while at the same time minimizing glare. In considering this problem there are two types of glare to be recognized. Direct glare refers to the effect of a light source within the visual field; specular glare refers to the effect of surfaces which reflect light coming from outside the visual field. Research has indicated that direct glare may be reduced by (a) avoiding bright light sources within 60 degrees of the center of the visual field, (b) using shields, hoods, and visors to keep direct light out of the viewer's eyes, (c) using indirect lighting, and (d) using several low-intensity lights instead of one high intensity light. Specular glare may be reduced by (a) using work surfaces and tools that diffuse reflected light, (b) using a diffused light source, and (c) positioning light sources and work so that light is not reflected toward the eye.

DISPLAY/CONTROL DESIGN FOR MONITORING TASKS

Monitoring type inspection tasks are typically performed at a console consisting of a set of displays and controls. The task of the inspector is to monitor the displays and, when appropriate, initiate action by means of one or more of the controls. Examples of inspection tasks performed in this context include performing functional tests on items of electronic equipment and monitoring an electrical switchboard to maintain proper electrical loads and to take action in emergency situations. In designing display/control consoles for tasks of this type there are two primary ways that inspection performance can be made most effective. These concern the layout of the displays and controls in the space allotted for them and the selection of the particular display and control elements used.

Layout of Displays and Controls

The problem encountered in locating displays and controls is that, usually, there are too many display and control elements and too little space in which to put them. Consider, for example, the preferable visual display area of a console. Guidelines for equipment design generally specify that the vertical range should extend from a point about level with the eyes to approximately a 45-degree angle below the horizontal line of sight. The preferred lateral angle range is also an angle of approximately 45 degrees from the line of sight. Within this area continuous visual attention is possible with normally comfortable changes in eye movement and relatively nominal head movements. If there are very many displays to be monitored, however, this area is too small to contain them all. Consequently, some guidelines are required for the determination of which displays and associated controls should be placed in this optimal area and which should be placed in more peripheral areas of the console.

There are three primary considerations for the placement of display/control elements. These are importance, functional relationship, and sequence of use. There are a number of reasons why one display or control may be considered more important than another. A more frequently used display is more important than one less frequently used. A display that signifies an emergency condition may be more important than a display used in a routine operation. A control requiring a high degree of speed and accuracy is more important than one that does not. In a consideration of the functional relationship of a set of displays and controls those that are related logically to an identifiable function should generally be grouped together; for example, a control used to regulate some condition should be located adjacent to the display that indicates the nature of the condition. The third consideration, sequence of use, concerns the location of display and control elements in a way that permits a logical and easy flow from one to the other in the performance of the task. In general, elements which are used together should be located together to minimize the added time required for searching and reaching. Obviously it is not uncommon for these three considerations to conflict with each other in the location of displays and controls for a specific application. In the event that conflicts do occur a compromise must be determined based on the specific needs of the system being designed.

A technique that can be very useful in determining the relative importance, functional relationships, and sequence of use for the control and display elements in a specific application is link analysis. Link analysis is primarily concerned with determining the frequency and sequence with

Figure 7.1 Gyro compass signal converter test set.

which display and control elements are used. The four basic steps in this procedure are (a) counting the number of times each element will be used, (b) counting the number of times pairs of elements will be used sequentially or concurrently, (c) rating the criticality of the link between each pair of elements, and (d) using the frequencies and criticalities to determine the importance of the links for console design. Link analysis may be conducted graphically using preliminary panel layouts or analytically through use of a computer. The use of a computer for link analysis has been described elsewhere [7]. An illustration of the application of graphic link analysis to the design of a console used in the functional test of items of electronic equipment is given below.

In an attempt to reduce operating time, training time, and the extent of human error in test operations graphic link analysis was applied as part of an effort to redesign the functional test set for part of a missile auto-navigation system. This console is shown before its redesign in Figure 7.1. The basic task descriptions required for the analysis were obtained from a preliminary test manual. Equipment panels, rather than the display/control elements themselves, were used in the analysis because the console was built primarily from commercially available units. The number of functional links was computed by counting the number of times each

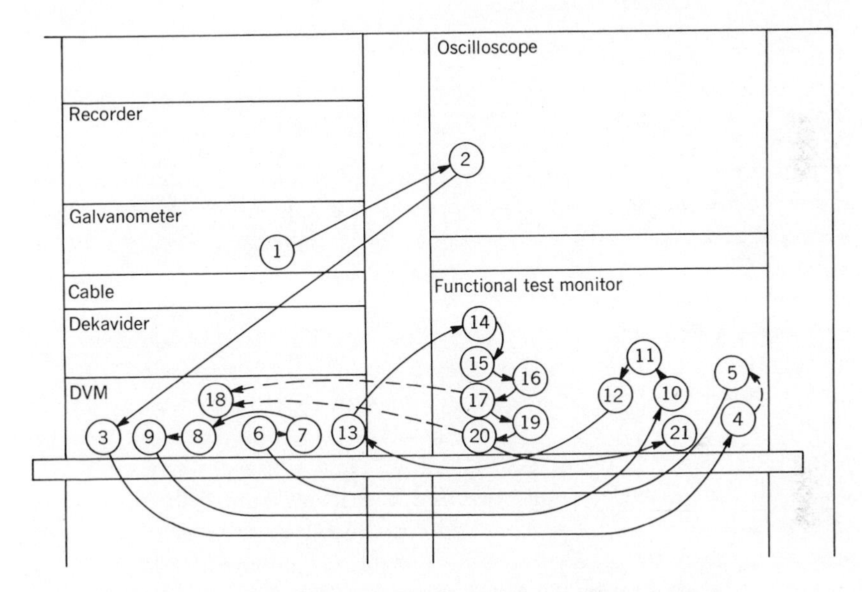

Figure 7.2 Link analysis of GCSC panel for sample test sequence.

panel was used; sequential links were determined by totaling the movements from one panel to another. For purposes of this analysis all panels were assumed to be equally critical. The graphic link analysis for one test sequence is illustrated in Figure 7.2.

The results of the graphic link analysis indicated that the existing panel layout was considerably less than optimum. The digital voltmeter, for example, accounted for more than half the number of visual links; yet it was located outside of the optimal visual display area, too low on the console for good viewing. Table 7.2 lists the number of visual and control links for each panel.

The analysis also indicated that the operators record sheet, functional

Table 7.2 Functional Links to Panels

Rank	Panel (or Item of Equipment)	Visual Links	Control Links	Total
1	Functional test monitor	1	116	117
2	Digital voltmeter	52	13	65
3	Operator's record sheets	—	62	62
4	Galvanometer	17	32	49
5	Voltage and temperature coefficient monitor	1	47	48
6	Oscilloscope	11	20	31
7	Recorder monitor	8	11	19
8	Temperature chamber	4	10	14
9	Dekavider	3	10	13
10	Cabel panel	—	3	3
	Total	97	324	421

test monitor, and the digital voltmeter had the greatest number of sequential links among them. Therefore they should be grouped as closely together as possible. The sequential links among the various panels are given in Table 7.3. As a result of the analysis a new arrangement of console panels was developed. The most frequently used display, the digital voltmeter, was placed in the optimum viewing position and the most frequently used panel, the functional test monitor, was placed directly below within the operator's optimum span of reach. The writing surface for the operator's record sheets remained in the same place but was now just below the functional test monitor. Thus these three panels were optimally placed and grouped together and, as a result of the new layout, the amount of hand travel required for the functional testing operation was reduced by 53 percent. The new panel arrangement is illustrated in Figure 7.3 with graphic link analysis for the test sequence previously shown in Figure 7.2.

Selection of Displays and Controls

These is an extensive amount of information available on the specific types of displays and controls that are optimal for different applications. This information has resulted primarily from research conducted by engineering psychologists on human performance under alternative display control configurations. Much of this information may be found in the previously cited handbooks, and comprehensive collections of this information exist as design criteria listed in military standards for equipment design. These standards are frequently included as requirements in contracts for military systems. One of the most complete and useful standards

Table 7.3 Sequential Links Among Panels

	Digital Voltmeter	Operator's Record Sheets	Galvanometer	Voltage and Temperature Coefficient Monitor	Oscilloscope	Record Monitor	Temperature Chamber	Dekavider	Cable Panel
Functional test monitor	47	41	8	1	15	0	0	0	0
Digital voltmeter		41	2	24	2	0	0	1	0
Operator's record sheets			4	1	16	12	1	4	0
Galvanometer				1	1	3	0	16	2
Voltage and temperature coefficient monitor					0	0	4	1	0
Oscilloscope						0	0	0	0
Record monitor							1	2	0
Temperature chamber								1	1
Dekavider									3

is Military Standard 803A, "Human Engineering Design Criteria for Aerospace Systems and Equipment," published in 1964 by the Department of the Air Force. In addition to specifications for workspace design and display/control configurations, design criteria are included for evaluating equipment maintainability, labeling, safety, and standardization.

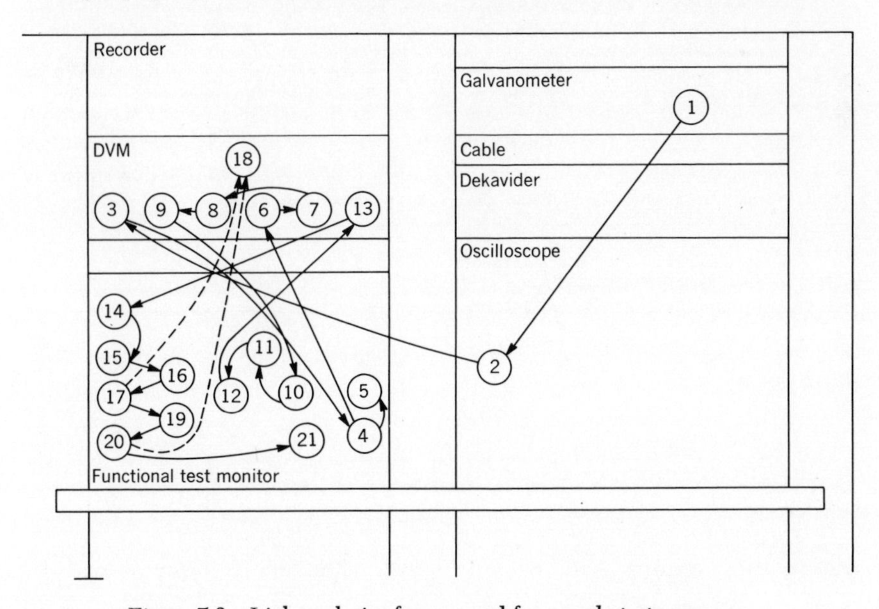

Figure 7.3 Link analysis of new panel for sample test sequence.

As a method of organizing and applying available information in the selection of displays and controls and in the specification of their arrangement on console panels, it is suggested that a checklist be developed for the specific console being designed. A review of the referenced documents may be completed to identify the design criteria that are applicable to the problem at hand. These, then, may be compiled into a checklist for use in evaluating preliminary design layouts and display/control selections. Table 7.4 illustrates one method of summarizing the results of a checklist evaluation of a display/control console. The part of the evaluation shown in this example concerns the labeling of displays and controls on the console panel. The summary is keyed to specific check list items and includes a description of the discrepancy, its impact, and recommendations for design changes. Obviously use of the checklist is not a guarantee that resourcefulness and imagination will be employed in the design of a display/control console. Its primary value is in assuring that available information on human performance will not be overlooked in the design effort.

Table 7.4 Example of a Human Engineering Evaluation Format

Checklist Reference	Description of Discrepancy	Impact	Recommendations
5.1.1	Rotary selector switches in the COMMAND FUNCTIONS and the MALFUNCTIONS areas are not labeled as to function	Slight—might be confusing to new operator in going from manual to console panel	No change—addition of labels would cause overcrowding of panel face
5.1.2	Labels for the ACTIVATE push-buttons might be confused with the rotary selector switches above them	Moderate—the operator might easily confuse the two controls, particularly as the rotary switches are not labeled	Outline the ACTIVATE labels and push-button areas with $\frac{1}{16}$ inch black lines
5.1.4	The abbreviations "XI" and "X2" on COMMAND FUNCTION rotary switch do not clearly imply "Target 1" and "Target 2," nor do they coincide with the abbreviations used in the manual	Moderate—likely to lead to misinterpretation by operator	Change abbreviations to read "TGT 1" and "TGT 2"

SUMMARY

1. In any organization of inspection work an attempt should be made to minimize the number of quality characteristics considered at each inspection station, within the limitations of costs and available personnel.

2. Inspection tasks are demanding in terms of the effort and concentration they require and, consequently, inspection stations should be

located where the amount of disturbing influence and the potential for unrealistic time pressures are minimized.

3. Consistent with the objective of minimizing inspection task complexity, the number of different quality standards used by any given inspector at any given time should be kept to a minimum.

4. Inspection performance may be enhanced by organizing inspection work in a manner that will permit the use of the most effective tools and techniques. Since the tools and techniques employed will have an impact on the manner in which personnel are assigned, quality standards defined, and inspectors instructed, tools cannot be considered independently of these other aspects of the inspection job.

5. The only communications normally considered in the inspection job are those associated with providing information such as quality standards and procedural instructions to the inspector and obtaining quality data from the inspector. Frequently overlooked are communications associated with inspector motivation. Among the important motivational communications are those that permit inspectors to suggest ways of reorganizing inspection work, those used to establish inspection goals, and those that provide information on inspection performance.

6. Written inspection instructions should include one or more of the following six basic types of information needed by the inspector: when to inspect, what to inspect, inspection standards, inspection methods, how to report data, material handling, and administrative procedures.

7. Among the guidelines which may be used to assure the effectiveness of written inspection instructions are the following: establishing a specific objective for each instruction, stating the instruction in operational terms, using only necessary, common words, and eliminating the use of adjectives and adverbs.

8. As a basis for considering both the activities to be performed and the characteristics of the inspectors in designing the inspection workspace, detailed information should be obtained on the tasks to be performed. In light of this information the general nature and environment of the workspace required and the workspace dimensions can be determined. Human factors engineering handbooks are available to provide workspace dimensions appropriate for specified work conditions and segments of the worker population.

9. The primary visual consideration in designing an inspection workspace is the illumination and how it is provided. The two aspects of illumination to be considered are illumination level and light distribution.

10. The three primary considerations for the placement of display control elements in designing a monitoring type task are importance, functional relationship, and sequence of use. Link analysis is a useful

technique to employ in designing a display/control panel in light of these considerations.

11. To make the best use of the extensive amount of information available on the specific types of displays and controls it is recommended that a checklist be developed for each display/control design problem.

REFERENCES

[1] Edward C. Schleh, *A study of Human Factors Affecting Quality and Reliability in Unmanned Spacecraft Components.* Minneapolis: Schleh Associates, 1966.

[2] Wesley E. Woodson and Donald Conover, *Human Engineering Guide for Equipment Designers* 2nd Ed., Berkeley: University of California Press, 1964.

[3] Henry Dreyfuss, *The Measure of Man.* New York: Whitney, 1960.

[4] Clifford T. Morgan, et al. (Eds.), *Human Engineering Guide to Equipment Design.* New York: McGraw-Hill, 1963.

[5] Richard H. Blackwell, Development and Use of a Quantitative Method for Specification of Interior Illumination Levels on the Basis of Performance Data. *Illum. Engr.*, 1959, **54**, 317–353.

[6] C. L. Crouch, New Method of Determining Illumination Required for Tasks. *Illum. Engr.*, 1958, **53**, 416–422.

[7] Robert C. Haygood, Kenneth S. Teel, and Charles P. Greening, Link Analysis by Computer. *Human Factors,* 1964, **6**, 63–70.

VIII

MEASURING INSPECTION PERFORMANCE

As one of the three key elements in the scientific method, measurement plays a significant role in evaluating and improving inspection operations. In spite of its importance, however, the measurement of inspection performance is frequently given little attention or is completely ignored. While it is unheard of to design and produce even a simple appliance or piece of equipment without subjecting it to a series of performance tests, inspection jobs are usually designed and personnel selected and trained to fill them without a thought being given to determining how well this inspection system works. Furthermore, in those cases in which adequate measures have been made of inspection performance, the performance of the job/inspector is usually found to be unsatisfactory. This, of course, is no disgrace because a newly designed instrument, appliance, or automobile seldom passes its first performance test either.

The importance of inspection accuracy in meeting quality assurance objectives indicates that every attempt should be made to obtain satisfactory measures of inspection performance. Performance measures should be emphasized early in the production program for purposes of evaluating and changing the design of the inspection job. In addition, provision should be made for periodic measures of inspection performance throughout the period of production for purposes of identifying weaknesses in the inspection job or in the inspectors who are performing the inspection work. The resulting information will provide a sound basis for making improvements in the various aspects of the inspection job such as the design of the inspection station, the tools and methods provided, and the inspection instructions. These data may also provide a basis for identifying personnel needs in terms of requirements for training, selection, or motivation.

The inspection supervisor's job is one of establishing the proper conditions for meeting quality objectives through the activities of planning, decision making, and communicating. Critical to the satisfactory performance of these activities are measures of inspection performance. Adequate measures of inspection accuracy permit the supervisor to plan, make decisions, and communicate on the basis of fact rather than guesswork and provide him with the insights that lead to the solution of inspection problems. In addition, these measures provide a sound basis for the supervisor's evaluation of the effects of the changes he has made in his operations.

CHARACTERISTICS OF A SATISFACTORY MEASURE

There are several different methods that may be employed in the measurement of inspection performance. In selecting a satisfactory measure for a given set of circumstances there are four primary considerations. These are the extent to which the measure is relevant, consistent, unbiased, and workable.

The relevance of a performance measure refers to the extent to which it is meaningful in terms of the objectives for which it is used. For example, if the performance of a sample of inspectors is being measured to determine how well the inspection job has been designed, the method of measuring performance should give a great deal of weight to aspects of inspection accuracy. A method that provided data on only the speed with which inspections were performed without considering the percentage of defects detected would not be a relevant one. To be relevant a measure must typically include the critical aspects of the job, that is, the things that are done that really make a difference.

To be consistent a measure must be capable of producing approximately the same results from one time to another under similar conditions. This consistency or stability is frequently referred to as the reliability of a performance measure. A procedure that resulted in relatively little relationship between measures from one time to another would be termed unreliable. A common method of estimating the reliability of a measure is by comparing performance values for a sample of individuals obtained at two different periods of time. A correlation coefficient indicating the relationship between the two sets of measures would provide an index of reliability. Such an index would have a range of .00, indicating zero correlation between the measures, to 1.00, indicating the two sets of measures to be identical. In general, a reliability index above .85 would indicate a highly reliable measure and an index above .75 would indicate an acceptably reliable measure for most purposes.

To be unbiased a measure must be free of factors that would distort (bias) the results. As an illustration of how bias can operate, consider measuring the performance of a sample of inspectors to determine the types and extent of training required. If inspection performance was measured on items that were seldom encountered in their jobs, the resulting performance data would be contaminated by an irrelevant factor and the measures obtained would not be very useful for the identification of specific job training needs. In general, bias can be minimized simply by assuring that the measurement method imposes the same conditions normally encountered on the job for all persons being measured.

A fourth consideration in developing or selecting a performance measure is the practical aspect of being able to apply it without great difficulty or cost. There is usually a trade-off between the quality of the results and the ease of use. In the measurement of inspection performance, as it is in many things we do, the methods that yield the greatest quality are also the measures that are most costly and difficult to apply. Obviously, if two measures will provide similar results, the easier one to use would be selected. On the other hand, there is no good reason for using an easily applied measure if it does not meet adequate standards of relevancy, consistency, and freedom from bias. In the following section several types of inspection performance measure are described and considerations in their use discussed. These measures include the inspection job sample, the repeated inspection, the inspection audit, and supervisor ratings of inspection performance.

THE INSPECTION JOB SAMPLE

The inspection job sample is a method of measuring inspection performance which has been shown to be relevant, consistent and unbiased. In its simplest form an inspection job sample involves inspections of items which have defects that are known to the supervisor or whoever administers the job sample but unknown to the inspectors whose performance is being measured. As the name implies, the job sample is really just a sample of the work regularly performed by a group of inspectors. It has the added feature of providing the detailed information about inspection performance that cannot typically be obtained directly from the job. Job sample performance measures may be adapted for use with measurement, scanning and monitoring type inspection tasks.

Inspection job samples should be used primarily to explore how well the inspection job has been designed and how well the supervisor is doing in developing his people and in providing them with the information, tools, aids, and techniques they need in performing their work. Measuring the

proficiency of individual inspectors is only useful as a means of identifying weak areas that may be improved through training and changes in the job. Job sample performance measures should never be used as the basis for disciplinary action. Disciplinary action would be inappropriate because a low level of inspection performance may be as much the fault of the supervisor as of the inspector. Unless the supervisor has provided adequate training, procedures, and other job requirements, the inspector cannot be expected to demonstrate an acceptable level of inspection performance.

Inspection job samples have been employed successfully by supervisors, quality engineers, and human factors specialists to improve performance in a wide variety of inspection jobs including the inspection of electronic chassis, modules, printed circuit boards, photographic materials, precision instruments, paperwork, machined parts, tin plate, television picture tubes, bottles, microelectronic circuits, and many other products. Job samples have provided the means for measuring initial performance levels, identifying problems, and evaluating the impact on performance of alternative problem solutions. In addition, they have been useful in developing a better understanding of the nature of inspection jobs and thus providing better insights into the use of inspection personnel. An illustration of the basic aspects of the job sample approach is shown in Figure 8.1, and the steps involved in developing and using inspection job samples are provided.

There are two primary limitations in the use of job samples. As the name implies, the job sample only samples performance and as a result may not have a perfect correspondence with everyday job behavior. Motivational factors, for example, may cause the job sample to differ somewhat from daily performance on the job. As a consequence the job sample is better used for determining job/inspector capabilities than it is for estimating day-to-day on-the-job inspection performance. A job sample may require somewhat more time and effort to develop and use

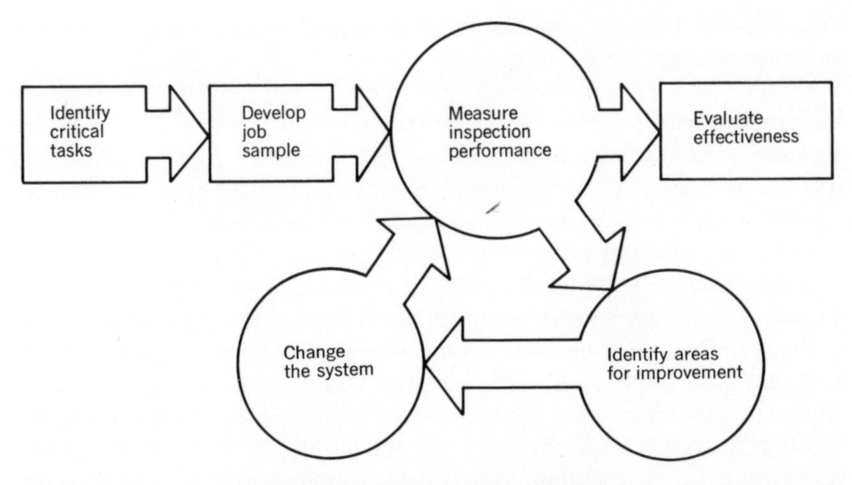

Figure 8.1 The job sample approach.

than certain other measures of performance such as supervisor judgments. Therefore a degree of care should be exercised in initiating the use of job sample performance measures. They should be used only for those objectives for which they are most appropriate. The following steps are listed as guidelines in the development and use of inspection job samples. Although these steps are oriented primarily with respect to scanning type inspection tasks, they are easily adapted to measurement and monitoring type tasks.

1. *Select a set of items to be inspected.* From the various items regularly inspected, select two or more sample items. The items may be obtained from scrap or they may be taken directly from production. An attempt should be made to obtain items which have a representative set of defects, or to build into them representative defects. Enough items should be obtained so that the set contains enough defects for a meaningful statistical analysis. Thirty defects or more should be adequate.

2. *Develop a master defect list.* A panel of experts should determine the defects present in the items selected and should develop a master defect list. The panel should be organized to include people who possess knowledge about the relevant inspection standards for these items and who represent any potentially different points of view concerning defects in items of these types. It is suggested that the panel be made up of the inspection supervisor or leadman, the responsible quality control engineer, and an audit inspector. Each panel member should independently inspect each item, describe each defect found, and record its classification. After each panel member has inspected each item and recorded the de-

fects the panel should jointly review the defects and make a final determination of a master list of defects.

3. *Have each inspector inspect each item.* Each inspector should inspect each item independently. To ensure that the best working conditions are provided and that the inspectors are not subjected to interruptions, a space such as an office or conference room should be set up away from the normal work area. In this space all the reference materials and tools normally used to inspect the items should be immediately available. Each inspector should be given the same instructions. The instructions should include what the inspector should do and how his performance will be evaluated. The job sample instructions which were given to printed circuit inspectors are cited as an example: "You will be inspecting nine printed circuits, three from each of three inspection stations—detail, assembly, and finish. Inspect each circuit for all defects as defined by the reference instructions for that station. You will be supplied with all the specifications, tools, and prints needed to make your inspections.

"Before starting your inspections, you may have up to 15 minutes to review the reference materials. However, you may start to inspect any time you wish.

"Inspect each circuit as you would normally and record the defects and resistance readings on the sheet provided. Put the code number of the defect in the margin and draw a line to the defect location on the picture.

"For resistors that are out of tolerance record the actual reading obtained. Your performance will be evaluated on the basis of four measures:

(a) The percentage of true defects found.
(b) The number of defects recorded that are not true defects.
(c) The percentage of true defects reported that are correctly coded.
(d) The amount of time taken.

"Before using any equipment, check its setup. Since others will be inspecting these circuits, please be careful with them and do not discuss your findings with anyone. If any damage is done during your inspections, please point this out to me.

"Please write your name on all papers.

"Have you any questions?"

4. *Organize the data collected.* To permit efficient collection and analysis of performance data, a data collection matrix should be used. A separate matrix should be used for each item inspected. An example of such a matrix is shown in Figure 8.2. As can be seen, the master defect list generated by the panel is in the left-hand column. In the next column to the right the agreed-on classification for each defect is provided. Across the top of the matrix each inspector participating in the study is listed by

The Detection/Classification Made by Each Inspector for Each Defect in Module Assembly 1

Defect in Assembly 1	Defect Class.	Inspector 1	2	3	4	5	6	7	8	9	10	11	12	13	14	15	16	17	18	19
1. No inspection acceptance of bracket	C																			
2. Missing wire	A	A		A	A		A	A		A	A	A	A	B	A		A	A	A	
3. Voids in black poly on hardware	C		C														B			
4. Harness wires too long	C						C													
5. Illegible part number	C							C												
6. Solder splatter on black poly	C									C									C	
7. Improper wetting of solder	C																			
8. Loose support	C	C	C		C	C	C	C	C	C	C	C	B		B		B	B	C	
9. Reverse polarity	A		A		A	A	A	A	A	A	A	A	A	A	A		A		A	A
10. Lifted eyelet pad	C																		B	
11. Damaged lead	B						A				B									
12. Loose screw	B	C			C			C			B	C	B	C	B			B		
13. Solder splatter by resistor	C				B															
14. Solder splatter by resistor	C				B			B												
15. Pin hole solder void	C								C							B	B			
16. Raised eyelet pad	B		B		B	B	B				B	B	B	B			B	B		
17. Solder splatter by resistor	C				B															
18. Wrong parts (resistors)	A			A	A	A		A	A	A		A	A	A	A	A	A		A	A
19. Improper wetting of solder	C									C										
20. Contact strip protector missing	C	C		C	C		B		C	C			C				C			
Number of False Reports		8	2	2	10	4	12	10	3	22	2	1	1	3	2	17	47	6	4	10

Figure 8.2 Sample data collection matrix.

number code. In recording the data collected the classification of each defect detected by each participant is placed in the appropriate cell. Blank cells indicate that the defect was not detected. The last item in the left-hand column—False Defects—refers to defects reported which were not on the master defect list. The number of false defects reported is entered in the cell corresponding to each inspector.

5. *Compute the percentage of defects detected.* The data collection matrix provides the information required for computing indexes of performance. The percentage of defects detected is computed for each inspector by dividing the number of defects on the master list into the number of master list defects detected and multiplying by 100. An inspector who detected 19 defects out of the 30 on the master list would have a detection percentage of $19/30 \times 100 = 63$. A group of five inspectors who detected 19, 15, 25, 12 and 20 defects in the same items would have an average detection percentage of

$$(19 + 15 + 25 + 12 + 20)/(30 \times 5) \times 100 = 91/50 \times 100 = 61.$$

6. *Compute the percentage of false reports.* The percentage of false reports is computed by dividing the total number of defects reported (defects on master list plus defects not on master list) into the number of defects reported which were not on the master list and multiplying by 100. An inspector who reported a total of 37 defects, 20 of which were on the master list and 17 were not, would have a false report percentage of $17/37 \times 100 = 46$.

7. *Compute percentage of defects properly classified.* Of the master list defects reported, determine the number correctly classified. The percentage properly classified is computed by dividing the number of master list defects detected into the number correctly classified and multiplying by 100; for example, if 20 master list defects were detected and 14 were properly classified, the percentage properly classified would be 70.

8. *Identify problem characteristics.* Categorize the master list defects into logical groupings. Then, for each category, compute the average percentage of defects detected in this category. For example, if the number of defects in Category G is 5 and the number of defects detected within this category by all 12 inspectors is 55, then the average percentage of Category G defects detected would be $55/(5 \times 12) \times 100 = 92$. Finding by a similar computation that the percentage of Category H defects detected was only 46, we may conclude that the greatest emphasis should be given to improving the detection of Category H defects. Similar analyses would be conducted to identify the characteristics involved in the greatest number of false reports and to identify problem characteristics with respect to classification performance. Examples of two analyses

Percentage of Defects Detected Within Each Defect and CPC Category

Defect Category	CPC Category Detail	Assembly	Finish	Total
Conductor	97	33	89	78
Resistor	66	*	*	66
Solder, masks, joints	83	59	40	66
Leads, terminals	79	73	55	73
Contamination	17	54	55	47
Substrate	**	55	38	48
Component	*	58	62	59
Identification	*	35	55	40
Conformal coating	*	*	40	40
Total	71	54	49	59

Percentage of False Reports Within Each Defect and CPC Category

Defect Category	CPC Category Detail	Assembly	Finish	Total
Conductor	5	33	0	9
Resistor	20	*	*	20
Solder, mask, joints	29	31	43	31
Leads, terminals	0	25	8	10
Contamination	79	33	18	37
Substrate	**	28	20	25
Component	*	3	0	2
Identification	*	0	14	5
Conformal coating	*	*	19	19
Total	19	23	15	19

* No defects of this category present
** Insufficient data to compute a percentage

Figure 8.3 Sample analysis to indentify problem characteristics.

of this type are shown in Figure 8.3. The circled percentages indicate the categories of characteristics which caused the greatest problems in circuit inspections.

9. *Initiate corrective action.* The final step involves selecting and properly executing the appropriate course of action to overcome the problems identified. Unless the decision is obvious, it would be best to plan the course of action in conjunction with appropriate specialists—quality control engineers, training specialists, and human factors specialists. Execution of the selected course of action can be aided by reviewing the appropriate chapters in this book.

THE REPEATED INSPECTION

The repeated inspection of an item or set of items may be a useful measurement method and is less costly to develop and use than an inspection job sample. This method is similar to the inspection job sample except that the items inspected do not contain a verified set of defects. Use of the repeated inspection will not result in estimates of inspection accuracy but will provide estimates of the amount of agreement among inspectors on the quality characteristics present; for example, if a set of items was selected and subjected to the inspections of each of a sample of inspectors, problem areas could be identified in terms of the characteristics about where there was a low level of agreement. As in the use of the job sample, the data would be analyzed to identify problems and appropriate corrective action would be taken.

THE INSPECTION AUDIT

Procedures which have been developed to audit performance on a group or departmental basis are typically special cases of the inspection job sample. The key aspect of an audit involves random reinspections of items by "audit" inspectors; for example, through the use of random selection techniques, a sample of about 10 percent of the items inspected during a week by the inspectors of a given department are reinspected by members of the audit organization. Defects that are found in the items and not detected by the inspectors are noted. These data are analyzed to provide general performance indexes such as Estimated Outgoing Quality Level and Inspection Accuracy Level.

There are two basic reasons why audit data, although necessary for determining the general status of quality and of the inspection operation, do not typically provide the detailed information required for effective problem solving. First, the audit inspector is often subjected to the same job environment as the inspector, a condition that may cause him to exhibit the same performance weaknesses as the inspector. As a consequence, the auditor's reinspection may overlook some of the same things that were overlooked in the original inspection and thus not shed any light on an underlying problem. Second, the nature of the data resulting from audit inspections is a function of the submitted quality level; for example, if perfect items are submitted to inspection for a given period, nothing could be learned from audit data about the inspection operation because there would be no opportunity to measure the effectiveness of inspectors. Only when inspectors have the opportunity to detect a representative sample of defects in the items audited would there be an adequate opportunity to measure inspection accuracy.

Inspection audits and job samples should be used as complementary measures. Job samples are probably too expensive to employ continuously. Therefore inspection audits are required to approximate the overall status of inspection performance and provide the basis for other useful indexes. When audits or other indications suggest that an unsatisfactory condition exists and, in any case, at regular intervals, the supervisor should employ inspection job samples to obtain in-depth information on the performance of his inspectors. In this manner he can anticipate problems and solve them before they become large and he can obtain the information needed to determine the nature of existing problems.

SUPERVISOR RATINGS

An alternative to the use of job samples and audits in the measurement of inspection performance is the use of supervisor ratings. In contrast to the more objective measures represented by job samples, supervisor ratings rely to a great extent on the judgment of individual supervisors. Therefore ratings may be biased by factors that are not related to performance. In addition, supervisors may not have an opportunity to observe certain critical activities such as the inspectors' ability to detect defects. Because they may not be relevant or free from bias, care must be exercised in the use of supervisor ratings for measuring inspection performance.

A study of machined parts inspectors compared the use of job samples with the use of supervisor ratings in measuring inspection performance. The job sample consisted of two machined parts, a metal bracket, and an electric motor support, developed to require a wide range of inspection techniques. A master list of the defects present in each part was developed by two lead inspectors and two engineering inspectors. There were about 100 characteristics to be inspected in each part, about one-third of which were defective. Each of the 26 experienced machined parts inspectors inspected each part. Four different aspects of inspection performance were measured for each type of characteristic: (a) detection—the percentage of defects identified, (b) identification and measurement—the percentage of defects found and correctly measured, (c) false detections —the percentage of reported discrepancies which were not out of tolerance, and (d) inspection time.

Supervisor rankings of the 26 inspectors were obtained from four inspection supervisors before the job sample performance measures were obtained. The inspectors were ranked in terms of overall job performance and, two weeks later, in terms of ability to detect defects. Each supervisor ranked only those inspectors whose work he had observed. The correlation between rankings on overall performance and rankings on detection ability was high (0.80). In other words, the supervisory rankings on these two aspects of performance were nearly the same. The supervisors either considered detection ability to be a major part of overall performance or they were simply responding to their "overall impression" of the inspectors in both instances.

The agreement between supervisor ratings of detection ability and job sample measures of detection performance was relatively low (0.37). This correlation was too small to indicate that ratings could be used in place of job samples to measure inspection capability.

Another contrast between supervisor ratings and job samples was in the amount of information that could be provided by each. While the only measure that could be made with ratings was some overall impression of inspection performance, the job sample made possible detailed data on four aspects of inspection performance for each of four types of quality characteristics. Consequently, although ratings gave almost no useful information in problem solving, the job sample measures yielded detailed data on strengths and weaknesses for most of the important aspects of machined parts inspection.

SUMMARY

1. Just as it is necessary to measure the performance of a new device before using it, it is necessary to measure the performance of inspectors in their jobs before using the inspection information produced. The importance of inspection accuracy in meeting quality assurance objectives indicates that every attempt should be made to obtain satisfactory measures of inspection performance on a regular basis.

2. The four primary considerations in selecting a satisfactory measure of inspection performance are relevance, consistency, freedom from bias, and workability. To be relevant a performance measure must be meaningful in terms of the objectives for which it is used. To be consistent a measure must be capable of producing approximately the same results from one time to another under similar conditions. To be unbiased a measure must be free of factors which would distort (bias) the results. To be workable a measure must be capable of being applied without great difficulty or cost.

3. The inspection job sample is a method of measuring inspection performance which has been shown to be relevant, consistent, and unbiased. Although a job sample may require somewhat more time and effort to

develop and use than other measures of performance, inspection job samples provide more information than other measures on the capability of the inspection job and the inspector.

4. Inspection audits are useful for determining overall inspection accuracy levels and general performance indexes. They are not as useful as job sample measures, however, for detailed problem solving. Inspection audits and job samples can be employed effectively together as complementary measures.

5. Supervisor ratings of inspection performance are generally unsatisfactory measures of inspection performance.

IX

INSPECTION TOOLS AND TECHNIQUES

A quality system meets its objectives through the effective interplay of people, equipment, and information. This interplay is well illustrated by the activities of the inspector; he does very little that does not involve some association with either information or equipment. For example, he interprets specifications (information), makes measurements and comparisons (usually involving equipment of some sort), and records the data he obtains (information again). The effectiveness of the inspector in performing these activities depends on the specific tools and techniques with which he works.

The effectiveness of inspectors is usually overrated. There is a common belief that an experienced, highly motivated inspector will usually detect between 80 and 100 percent of the defects present in a product regardless of the nature of the inspection task. However, when inspection performance has been measured under controlled conditions, percentages of defects detected have seldom been more than 80. Typical of these controlled studies were those conducted by Harris [1]. Using inspection job samples, performance in inspecting 10 different items of electronic and electromechanical equipment was measured. Each item was inspected only by individuals experienced in inspecting that item and currently assigned full time to inspection of items of that type. Each item contained a representative sample of defects; the defects had been verified and classified by inspection supervisors and quality engineers. In each case, inspections were conducted individually under relatively ideal working conditions. Work was performed in a quiet place away from the general work area, all tools and documents normally provided were available to the inspector, and each inspector had as much time as he wanted to make his inspection. But even under these conditions the average percentage of defects detected ranged from fewer than 30 for the most complex items to no greater than 80 for the least complex items. These results, generally consistent with findings from other studies conducted under controlled conditions, suggest that high levels of inspection performance should not be taken for granted. They also suggest that attention should be given to precisely how the inspection job is designed and to the specific tools and techniques provided to the inspector for his work.

New concepts in inspection tools and in methods employed by the inspector in his job have demonstrated a significant impact on inspection performance. Being human, inspectors have certain inherent limitations and capabilities with respect to inspection operations. The basic idea is to improve inspection performance by overcoming the inspector's human limitations and enhancing his human capabilities. The development and evaluation of several approaches to this objective are discussed in this section. The thinking behind each concept is provided and the approach taken to develop the concept into a workable tool or technique is discussed in some detail. By providing this information it is hoped that the reader will gain a familiarity with the methods of human factors research and engineering as well as the knowledge of some concepts that are applicable to the improvement of quality systems.

SCANNING METHODS

In scanning type inspection tasks the primary activity of inspectors is searching the product for defective conditions. Tasks of this type are difficult to perform even under the best of conditions and the lowest levels of task complexity. As a consequence, some attention has been given to improving the efficiency of scanning methods. For most products there is usually more than one way to scan for defects. Are there any guidelines to use in determining the best scanning method? Three studies which provide a partial answer to this question are described.

Inspecting Stationary Versus Moving Items

Research on visual acuity indicates that people can see a stationary object better than they can see a moving object. Therefore, inspection accuracy would be expected to be greater when scanning a product that is stationary than in scanning a product while it is moving. This conclusion is supported by the results of a study of inspection performance under actual shop conditions. The study investigated methods used in the job of tin plate inspection [2]. This inspection operation consisted of scanning sheets of tin plate for defects in appearance. The inspection was made in the process of turning the sheets from one stack to another. For each sheet the inspector determined from the appearance and feel of the sheet whether it was a prime or second or contained any of a number of possible defects.

In attempting to determine the best method of inspection, an inspection job sample of 150 sheets was developed. About 60 of these sheets were perfect and the other 90 had defects of one type or another. Using this job sample, the inspection accuracy of 150 tin plate inspectors was measured. Then motion pictures were taken at 1000 frames per minute of 12 inspectors selected from those who had taken the job sample test. In this group were two inspectors who had been rated fast and accurate, two rated fast but inaccurate, two rated slow and accurate, two rated slow and inaccurate and, four rated average in both speed and accuracy according to the performance measures. The pictures were taken in the inspection area while the inspectors were working under normal conditions.

An analysis of the films revealed that the least accurate inspectors followed the movement of the sheets with their eyes as they were turned from one stack to another; that is, the inspections were being made while the sheets were moving. The most accurate inspectors, however, inspected the sheets during the time they lay on top of each stack. One side was inspected as it lay on the top of one stack before being turned; the other side was inspected after the sheet was turned. Attention was thus alternated between the two stacks of tin plate, ignoring the sheet while it was being turned. As a result of this finding, all inspectors were trained in the method used by the most accurate inspectors.

Should it be necessary to conduct a scanning type inspection of a moving product or products such as those which are passing on a conveyor belt, the best method seems to be to have the products move laterally past the inspector rather than move toward him. Evidence to support this idea was provided by the results of a study of methods for inspecting glass spheres [3]. In this experiment spheres were inspected for appearance under two test conditions—spheres moving toward the inspector and spheres moving laterally past the inspector. The job of the inspector was to identify those spheres with defective conditions. The results showed significantly greater accuracy when these objects moved laterally past the inspector than when they moved toward him.

Inspecting for One Characteristic at a Time

When inspecting complex products, the inspector is frequently required to scan the product for a relatively large number of different types of defects. How should the inspector approach this task? Should he divide the product into relatively small areas and look for all possible defects in one area before going on to the next, or should he inspect the entire

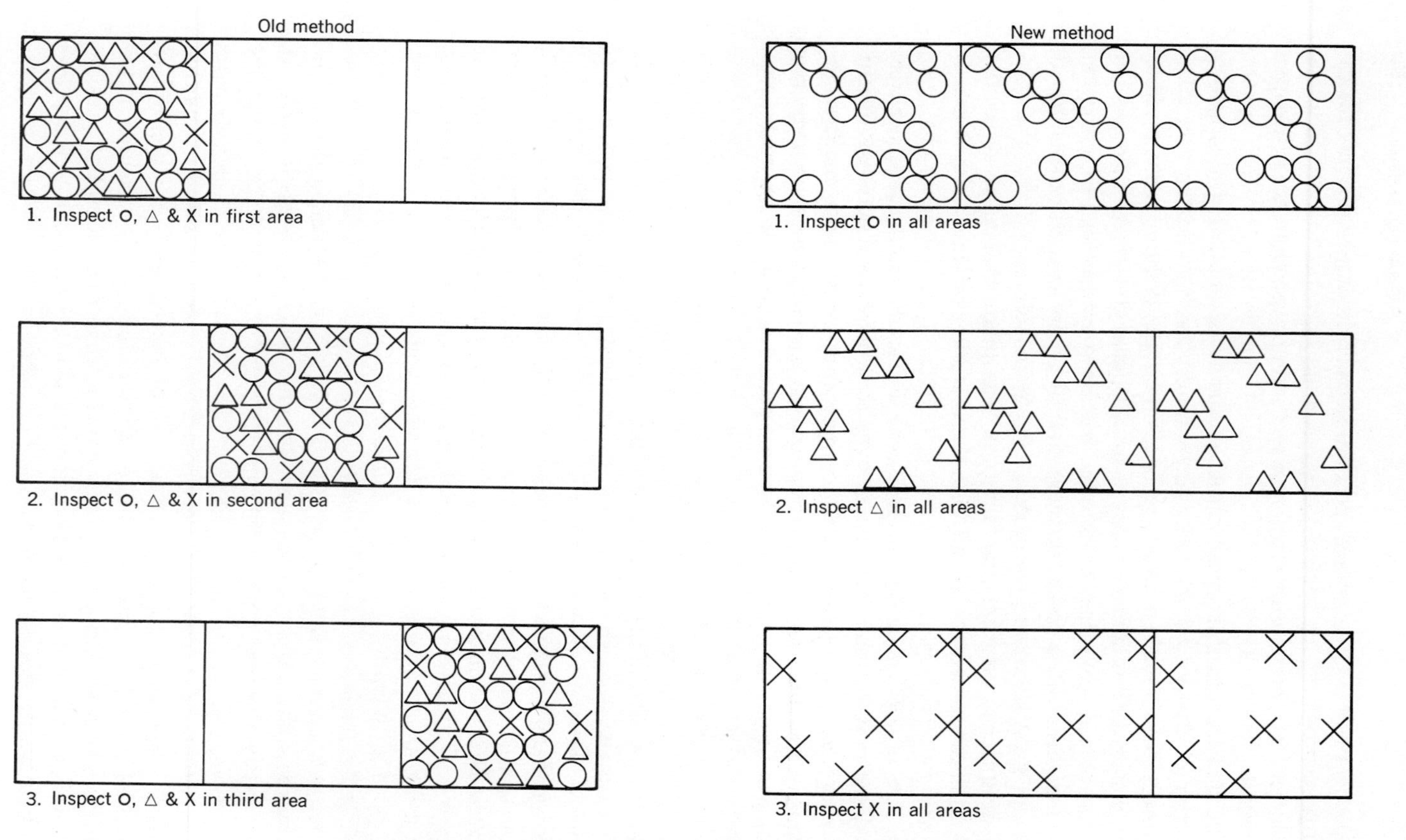

Figure 9.1 Symbolic comparison of old and new inspection methods.

product for one type of defect and then inspect the entire product for the next type of defect? To answer these questions an experimental study was conducted of techniques employed for the inspection of a complex electronic chassis [4].

The first step in this investigation consisted of observing a sample of experienced inspectors at work. These observations indicated that all of the inspectors observed used the area-by-area scanning method. Since there were over 25 different types of quality characteristic to be inspected, this meant that more than 25 different quality standards had to be applied simultaneously. This task would appear to be beyond the capability of even the most effective inspector.

A new characteristic-by-characteristic scanning method was developed and its effectiveness compared with the existing area-by-area method. In the new method inspectors attended to only one type of quality characteristic at a time. Inspection for each type was performed throughout the item before inspection for another type of defect was initiated. In this manner it was hoped to reduce the amount of "mental gear shifting" required of the inspector. Since visual inspection is primarily a matter of comparing the inspected characteristic with a mental image that corresponds with an acceptable characteristic, it was theorized that inspection accuracy would be improved by reducing the number, frequency, and variety of mental images required at any given time. The new method is contrasted with the old method in Figure 9.1.

To test the effectiveness of the new method, job sample performance measures were obtained of a sample of experienced inspectors when they used the new method and when they used the old method. The job sample consisted of two complex electronic chassis of the type regularly inspected by the inspectors. Half the inspector sample inspected one chassis with the old method and the other chassis with the new method. The other half of the sample did the same only in reverse order. Figure 9.2 shows an inspection being performed on the type of electronic chassis used in the study.

The new method was found to be 90 percent more effective than the old in detecting the critical defects present in the products. In addition, the characteristic-by-characteristic method resulted in more complete inspections. When the existing scanning method was used, certain types of defects were completely overlooked and some inspectors reported much higher percentages of certain defect categories than they did of others. In other words, being unable to attend to all of the quality characteristics and their associated standards, the inspectors attention was reduced to relatively few types of characteristics. Under the new method this bias was nearly eliminated.

Level of Magnification Versus Degree of Specialization

Many scanning inspection tasks require magnification of the object inspected. As the magnification level is increased, however, other things happen to the image of the object; for example, the visual field, object brightness, and the contrast between the object inspected and its background are all typically reduced. As a result, inspections involving several quality characteristics could actually be less effective under higher magnification levels.

Figure 9.2 Inspection of an electronic chassis.

Increasing the degree of specialization has been proposed as a means of gaining the most benefit from higher levels of magnification. Increased specialization is related to the characteristic-by-characteristic scanning method in that the inspectors' attention at any time is focused on just a few possible defects. A study of the combined effects of magnification level and degree of specialization on inspection accuracy was conducted at North American Rockwell. The inspection operations studied involved the inspection of components in microelectronic computers.

Inspection accuracy in detecting component defects was measured under two levels of inspection specialization and three levels of magnification. Experienced inspectors each inspected three computer modules with known component defects. Each inspector inspected the first module under 7X, the second under 15X and the third under 30X magnification; half the sample was instructed to inspect for only component defects and the other half to inspect for all possible defects. The order of presentation was counter balanced so that each board was inspected once under each combination of magnification level and degree of specialization. Inspectors were allowed as much time as they needed to complete their inspections.

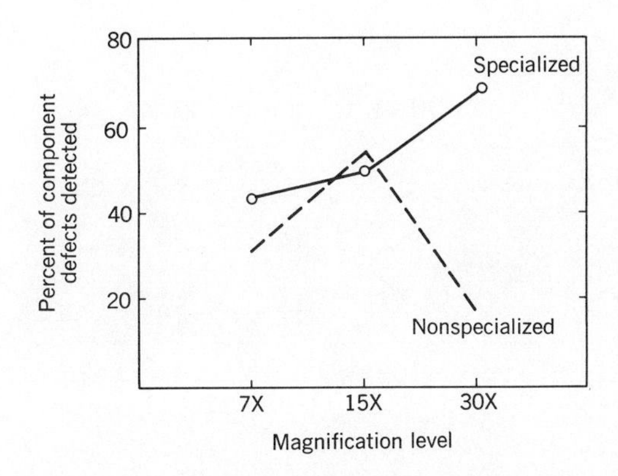

Figure 9.3 The effect of specialization and magnification level on the detection of component defects.

The findings indicated that the value of increased magnification is dependent upon the degree of specialization with which inspections are conducted. Inspection accuracy increased somewhat for both the specialization group and the nonspecialization group as magnification level was increased from 7X to 15X. Under 30X magnification, however, the group instructed to inspect for only component defects increased in accuracy but the group instructed to inspect for all defects decreased in accuracy. Thus, the highest level of inspection accuracy (68 percent) was under the specialization condition at 30X; however, the lowest level of inspection accuracy (16 percent) was also under 30X when inspectors were asked to inspect for all possible defects. This difference in inspection performance was statistically significant beyond the 0.05 level. These results are shown in Figure 9.3. It was concluded that the level of magnification for inspection tasks should be established in light of scanning procedures. Increasing magnification beyond a certain point without changing inspection procedures to accommodate for the new conditions may actually reduce inspection accuracy rather than increase it.

OVERLAY TOOLS

Some inspection tasks require detection of defects in a large array of similar items. Inspection of relative positions of pins in large electrical connectors and the inspection of electrical circuitry on circuit boards are examples of this type of inspection task. These tasks are nearly impossible

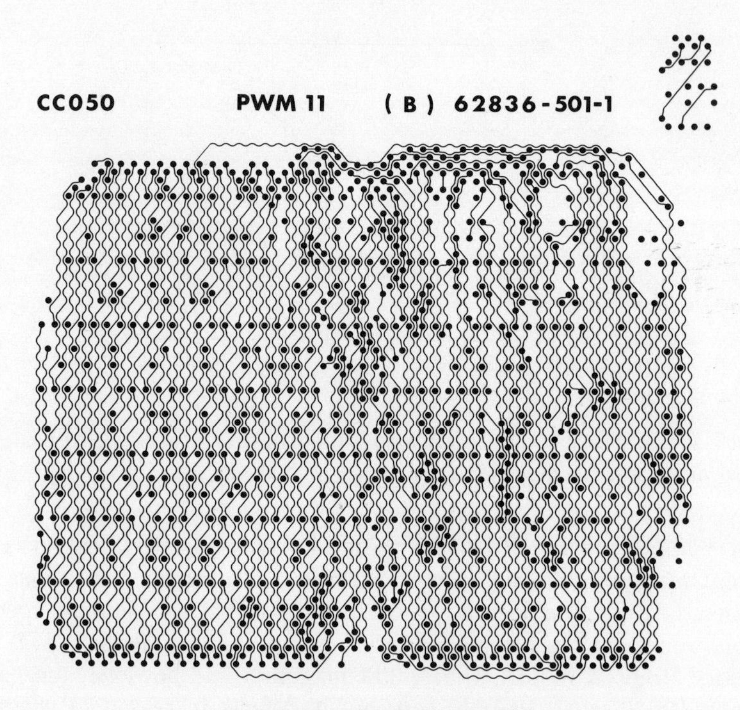

Figure 9.4 Sample photomask.

unless some frame of reference is used that will cause defective items to be highlighted. The size of this problem is illustrated by a typical layer of circuitry in a multilayer circuit board used in microelectronic systems. Such a circuitry layer may consist of as many as 3000 small circular pads interconnected by lines. Circuitry layers are inspected for defective pads and lines; defects include pads and lines which are undersized, oversized, misplaced, or damaged. For this type of inspection, it is often possible to design overlay tools that, when employed, cause defects to "pop out" at the inspector. The inspection task thus becomes one of looking for things that are highlighted as opposed to looking at each and every individual item and measuring items which appear to have out of tolerance conditions.

Photomask Overlays

A study by Sadler [5] illustrated the development and experimental evaluation of an overlay inspection tool. Photographic masks were used extensively in the development and production of multilayer circuit

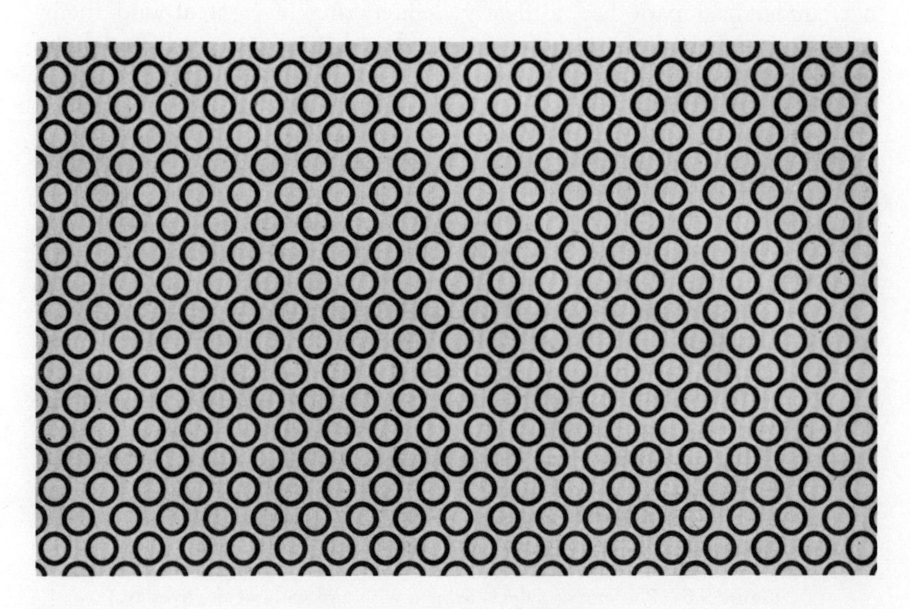

Figure 9.5 Photomask overlay.

boards used in special purpose computers. Because of the small tolerances involved and the high cost of rework or scrappage which can result from undetected defects, photomask inspections were both difficult and critical. An example of a photomask is shown in Figure 9.4. The initial step in this investigation was a study of the effectiveness of existing procedures for photomask inspection. Using existing procedures, an inspector scanned the photomasks for anything that appeared to him to be out of tolerance and then used a magnifying scale to measure the out-of-tolerance condition. A job sample consisting of eight masks with known defects was inspected by a sample of 20 inspectors. The results indicated that low percentages of defects were detected and high percentages of false detections were reported. Following this initial study, research was conducted to develop and experimentally evaluate overlay inspection techniques for these tasks.

An overlay inspection tool consisting of a matrix of gray rings on transparent material was developed. The gray rings corresponded to proper circuit pad positions. A section of the matrix is shown in Figure 9.5. By use of a set of guide pins for proper alignment the overlay was positioned over the photomask on a light table during the inspection task. Since the inner diameter of the rings was slightly larger than the maximum allowable pad diameter, pad defects could be detected in the following manner: undersized pads had a clearly defined ring of light around them, oversized pads had no ring of light at all and misplaced pads had light

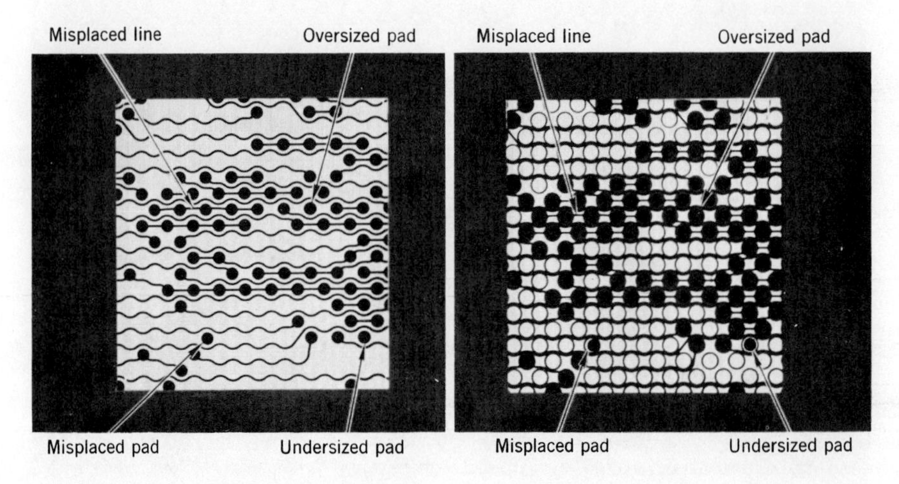

Figure 9.6 Photomask defects as seen with and without the overlay.

showing only on one side of the pad. Since the space between adjacent rings was approximately equal to the nominal line width, line defects could be detected in the following manner: undersized lines had light gaps between the lines and the rings, oversized lines overlapped the rings and misplaced lines either overlapped one ring or were too close to adjacent pads. Use of the inspection overlay is illustrated in Figure 9.6. Four defects are seen without the photomask overlay and with the photomask overlay.

Job sample measures of photomask inspection performance were used to evaluate the effectiveness of the inspection overlays. The job sample consisted of three photomasks containing a known set of out-of-tolerance pads and lines. Each of 30 inspectors individually inspected each master;

defects reported were compared with the master list of known defects and appropriate performance measures were computed. Effectiveness of the overlay technique was determined by comparing performance measures obtained when the overlay was used with measures obtained when the overlay was not used. Use of the overlay was found to result in significant improvements in inspection performance. An overall increase in inspection accuracy of 42 percent was obtained. As shown in Figure 9.7, however, the overlay was more effective with some defect categories than with others. Use of the overlay increased effectiveness in detecting oversized and undersized pads by over 250 percent but had little effect on

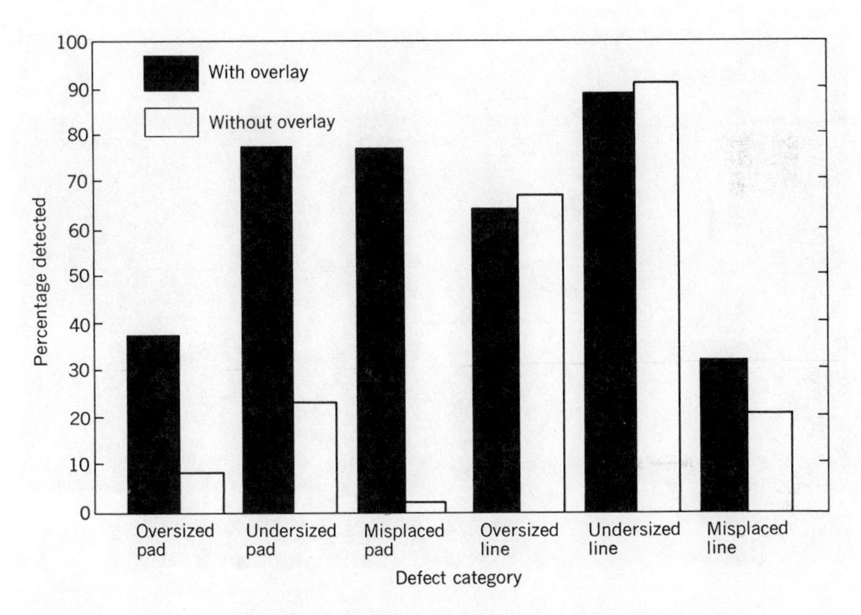

Figure 9.7 Effectiveness of overlays in inspecting photomasks.

detection of oversized and undersized lines. Apparently the visual cues available for the detection of undersized and oversized lines were sufficient without the use of the overlay. There was no significant difference between aided and unaided inspection in the amount of inspection time required.

Circuit Overlays

A second illustration of the use of inspection overlay methods was provided in a study of the inspection of circuit boards. At the time of the study, the primary source of quality problems in the fabrication of circuit boards for interconnecting microelectronic devices was the introduction of defects during the application and exposure of photoresistant material on the surface of the circuit board. The method being employed to detect defects at this stage of fabrication consisted of a visual search under 7X magnification with white light illumination for breaks and scratches in the blue photoresist material and for excess photoresist outside the lines of circuitry. The problems encountered with this inspection method included difficulty in detecting defects because of the fine tolerances required and the lack of cues for discriminating defective pads and lines from acceptable ones, false cues such as dust and surface imperfections

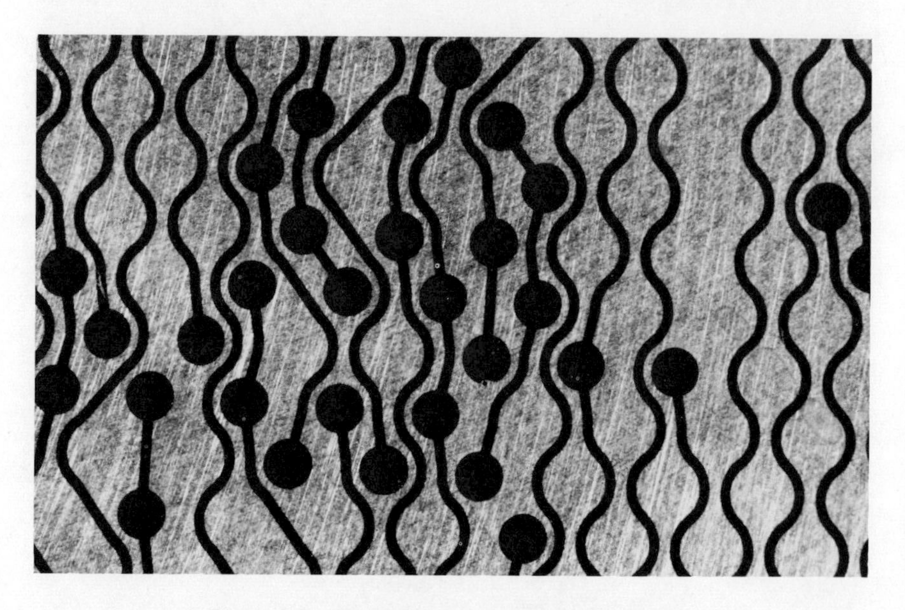

Figure 9.8 Circuit layer containing three defects.

leading to reports of defects where none actually existed, and glare of the light sources off the copper background.

A study was conducted to compare the effectiveness of the current inspection method with the effectiveness of using film overlays designed to minimize the three problems listed above. Two overlays were developed for use in inspecting each circuit. The idea behind the overlay was to increase the contrast between any given defect and its surrounding background; that is, to provide stronger visual cues for each defect. Since there were two basic types of defects which might be present—the lack of photoresist material where there should be some and the presence of photoresist materials where there should not be any—two overlays were required. A negative film overlay had the identical circuit pattern as the actual circuit but was black where the copper appeared on the circuit and was transparent where the blue lines and pads would appear on the actual circuit. Thus, when the film overlay was superimposed on the circuit any exposed copper in the blue circuitry would appear in sharp contrast to the general blue-black background. Figure 9.8 illustrates a section of circuitry containing three defects. Figure 9.9 illustrates the same section of circuitry with the negative film overlay super imposed over it. The defects appear as three light spots on the dark background.

Figure 9.9 Circuit layer with overlay exposing three defects.

To detect blue photoresist material where there should be none an orange positive overlay was used. This overlay, when superimposed on the actual circuit, covered the blue lines of circuitry with orange and, consequently, excess blue photoresist material would appear sharply against an orange-copper background. Excess photoresist material would be exposed in the manner shown in Figure 9.9 but would appear dark against a lighter background.

To evaluate the effectiveness of using the overlay films, a job sample consisting of four circuits was inspected by 12 inspectors—six used the current method and six used film overlays. Use of the film overlays resulted in a significant improvement in inspection accuracy over the current method. Of the defects present in the sample of circuits, 80 percent was detected using overlays and only 68 percent was detected using the current method. Use of the overlays resulted in greatest improvements in the detection of extra circuitry and in the detection of defective circuit pads. Improvements in excess of 30 percent occurred in these two defect categories.

COMPARISON TECHNIQUES

Because of the special capability of the human being to integrate what he sees, one way of increasing inspection effectiveness is to structure the inspection task so that the inspector makes direct comparisons between quality characteristics and quality standards presented to him simultaneously. Within reasonable limits of precision, most people can make accurate relative judgments concerning size, texture, color, condition, brightness, length, and so on. In addition, because human judgments of these types have been studied quite extensively, much is known about the relative precision of human judgments and how to measure this precision for specific applications. This area of psychological investigation is generally referred to as psychophysics; a number of the methods and findings have been discussed by Guilford [6].

Inspection for Density

In the inspection of raw materials, the standard for acceptance or rejection may be in terms of the extent to which an observable condition exists. Blemishes on the surface of the material may be acceptable up to a certain point and unacceptable beyond that point. This type of inspection task is illustrated by the preparation of silicon wafers for the production of microelectronic devices. Inspections are conducted to determine the extent to which wafers have crystal imperfections which disturb the energy conditions required for proper semiconductor behavior. Since the extent of crystal imperfections can be determined from the density of pits visible on the wafer surface, wafers are inspected for pit density. At the time of this study, methods for determining pit density involved counting the number of pits in several areas on the wafer while viewing the sampled area through a microscope. These counts were then multiplied by a conversion factor to arrive at an overall estimate of pit density for the wafer. If the pit density was determined to be beyond an acceptable level, the wafer was rejected. This procedure took as long as 45 minutes per wafer depending upon wafer size.

A new approach to determining pit density utilized a projection/comparison technique. The sampled area of the wafer was projected on a

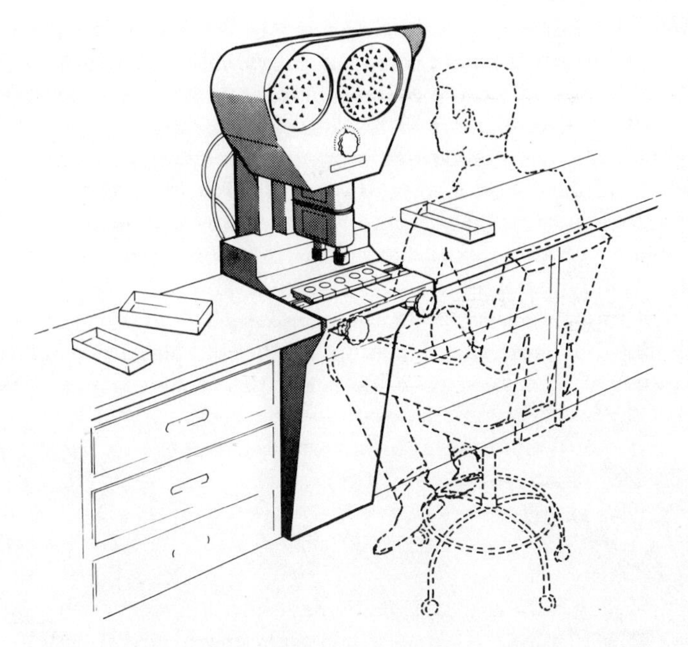

Figure 9.10 Projection/comparison system.

screen along with a standard. The pit density of the wafer was judged in comparison with the pit density of the standard. Wafers judged to have a greater density were rejected. An artist's conception of a projection/comparison system for inspecting wafer pit density is provided in Figure 9.10.

Before the design of a projection/comparison system was begun, a study was conducted to determine how accurately inspectors could discriminate between acceptable and unacceptable pit densities using projection/comparison methods. In addition, the relationship between inspection time and accuracy was investigated as well as the impact of training on inspector performance. A simulated projection/comparison system was developed. Eleven 35mm slides were prepared; each slide contained a random distribution of black dots. A slide with 200 dots was used as the standard; the 10 comparison slides provided increases from 150 to 250 dots in 10-dot steps. The projected area was 14 inches square and the dots were 1/16 inch in diameter. These materials were consistent with the configuration of available optics for projection/comparison systems and the acceptance standard currently in use. Figure 9.11 shows photographs of the simulated pits used in the study and actual pits produced under two different etch times used in silicon wafer preparation.

Sixty inspectors each made 44 comparative judgments of pit densities.

There were 11 different slides to be compared with the standard; each slide was rotated 90 degrees four times providing a total of 44 comparative judgments. Slides were presented in a random order. Each comparison slide was projected for 1, 5, or 10 seconds. Subjects had to decide whether the comparison slide had more or fewer dots than the standard and enter their answer on a score sheet within six seconds.

The results indicated that the projection/comparison technique can be used to make rapid and accurate comparisons of pit density. Accuracy was found to be unaffected by length of time allowed to make the comparison—performance was as accurate in 1 second as in 10. Inspection accuracy depended on the deviation of the sample density from the standard density. Accuracy was little better than chance for deviations as

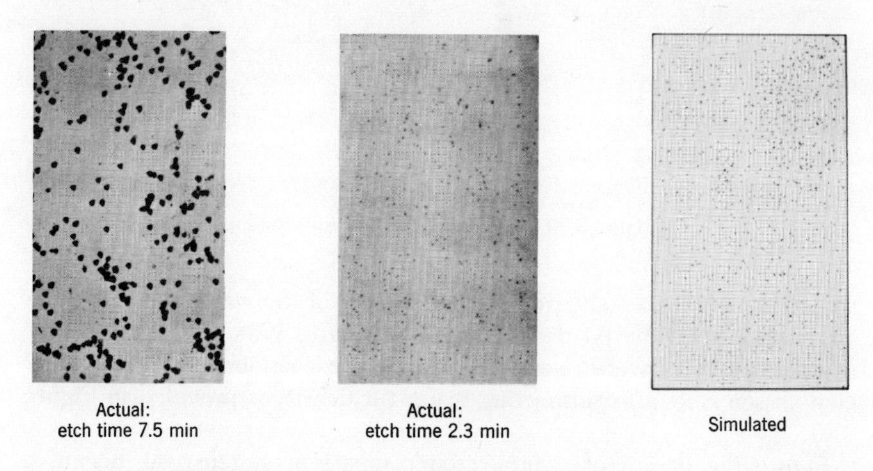

Figure 9.11 Actual and simulated wafer pit density.

small as 5 percent, whereas performance improved to 95 percent accuracy when deviations were as large as 20 percent. These results are shown in Figure 9.12. In practice a standard could be chosen to provide any desired level of accuracy in rejecting bad wafers; for example, selecting a standard with a pit density 20 percent less than the maximum acceptable density would result in correct rejections about 95 percent of the time.

Color Comparison

A second illustration of the value of comparative judgments in inspection operations was provided by the investigation of a problem in the fabrication of microelectronic devices [7]. During the fabrication of certain types of integrated circuits, a silicon wafer is coated with silicon dioxide. This oxide layer is 13,000 to 17,000 Å thick, depending on the device. During the formation of components this oxide layer is etched to various precisely calculated distances from the surface of the silicon substrate. If the desired thickness of the oxide from the silicon substrate is not held to a tolerance of ±200 Å, defects in the form of shorts and excessive voltages are likely to occur.

Because of the properties of silicon dioxide, one of the most precise and quickest ways of inspecting oxide thickness is making a determination of oxide color. As light entering the oxide layer reflects off the silicon substrate and passes back out of the oxide layer, light waves interfere with each other in a pattern determined by the thickness of the oxide. This alteration in light waves causes the light to appear colored; the color

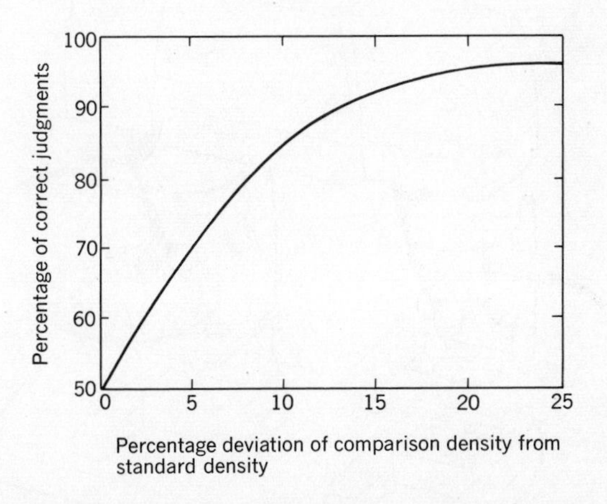

Figure 9.12 Inspection performance as a function of the deviation of comparison density from standard density.

perceived depends upon the thickness of the oxide. Initially, the standard employed for determining oxide thickness was a chart relating thickness in angstroms to a verbal description of the color. A preliminary investigation indicated that in using this method over 20 percent of etch thickness estimates were in excess of the 200 Å tolerance. As a consequence, an effort was undertaken to develop a more effective method for relating perceived color to oxide thickness.

Two alternative procedures were developed and evaluated. In one method the verbal color descriptions were replaced by actual colors. In this method the operator would (a) inspect the specimen wafer under the required magnification, (b) remember the color, (c) scan the set of actual color standards to find a color that matched the color of the specimen wafer, and (d) read off the thickness corresponding to the color standard selected. In the second alternative, a series of color standards would be available for simultaneous comparison with the specimen wafer. As a mechanism for providing direct color comparisons, a microscope such as the type used for simultaneous investigations in ballistics work, would be provided for this purpose. A ballistics microscope permits splitting the observed field into two parts. The wafer would be observed

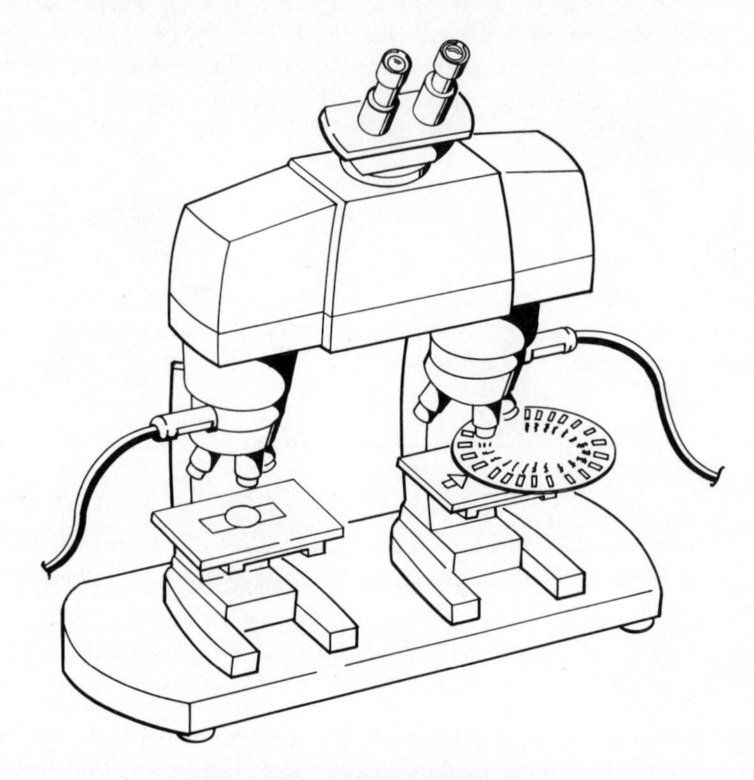

Figure 9.13 Direct comparison microscope.

in the left half of the field, for example, and the color standard in the right half of the field. A set of standards would be on a circular disc which could be rotated to select the desired standard. A microscope designed for this purpose is shown in Figure 9.13.

The relative precision of each of the three methods—verbal scale, color scale, and color matching—was determined. In this study color chips were employed to simulate actual materials so that the study could be conducted under the controlled conditions available in the laboratory. To simulate various oxide thicknesses, 50 color specimens were prepared by mounting color chips on a pink background. The pink background was similar to the pink color existing outside of the areas on the actual wafer. The 50 color chips covered a range of from 300 to 4000 Å units of etch thickness. Each of the three methods provided a means of determining etch thickness over this range in steps of 100 Å.

Direct color matching was found to be the most accurate technique. As shown in Figure 9.14, direct color matching resulted in a greater improvement in accuracy over the verbal scale than did use of the color scale. Direct color matching improved accuracy by 80 percent, whereas the improvement from use of the color scale was only about 29 percent. The practical significance of these results is illustrated in terms of the

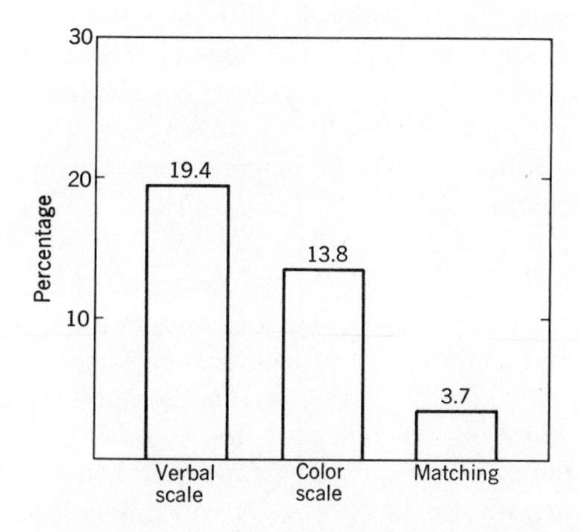

Figure 9.14 For each method, percent of responses exeeding the required accuracy.

potential reduction in production costs. It was estimated that approximately 15 percent of wafers containing etch errors greater than 200 Å will yield defective units. Assuming a production schedule of 2000 wafers per week at a production cost of $25 per wafer, use of the verbal scale method would result in a circuit loss amounting to $75,000 per year. This loss would be cut to $15,000 by employing the direct color matching method, a net gain of $60,000.

MAGNIFICATION

Some industrial inspection tasks obviously cannot be performed by the unaided eye. What is not so obvious, however, is the optimum level of magnification that should be provided for a given inspection task. There are several considerations involved in determining an optimum level of magnification; for example, inadequate magnification is likely to result in undetected defects. On the other hand, excessive levels of magnification may increase inspection costs as a result of the detection of minute defects which do not affect product functions. In addition, considerations such as field of view, depth of view, and distortion must be considered in selecting magnifying devices. To evaluate the impact of magnification on inspection performance accurately it is necessary to obtain actual performance data for the inspection tasks in question. Research has shown that the optimum magnification depends on the specific task as well as on the size of the object [8]. Consequently, if the inspection task is likely to have a significant impact on product quality and cost, an investment in determining optimum magnification levels may result in a sizable payoff. The recommended approach is to obtain measures of inspection performance under alternative levels of magnification. This approach was employed at North American Rockwell early in a major production program to determine the optimum level of magnification for use in inspecting ceramic printed circuits.

Establishing Level of Magnification

It was clear at the start of the production program for ceramic printed circuits (CPC) that some level of magnification would be required for the detection of critical defects in these assemblies; however, the effectiveness of various levels of magnification for these inspection tasks was unknown. Since the quality requirements for this program were high and the size of the program was large, a study was initiated to systematically establish magnification levels for inspecting these devices. Inspection performance was evaluated under five levels of magnification—unaided vision, 2X, 4X, 10X, and 20X. Eight different lots of CPCs with known defects were inspected by 10 experienced inspectors. Each lot of parts consisted of 10 CPCs and was examined by two different inspectors under each level of magnification. Therefore 800 CPC inspections were performed to detect a total of 654 defects. Inspection performance was measured in terms of the percentage of defects correctly detected. The relative size of a CPC under four of the magnification conditions is illustrated in Figure 9.15; the field of view and depth of field for each magnification condition are presented in Table 9.1.

Average defect detection performance under 10X and 20X was essentially the same; however, performance at 10X was significantly better than that at the three lower magnification levels. As shown in Figure 9.16, performance at 2X and 4X showed only small improvements over unaided

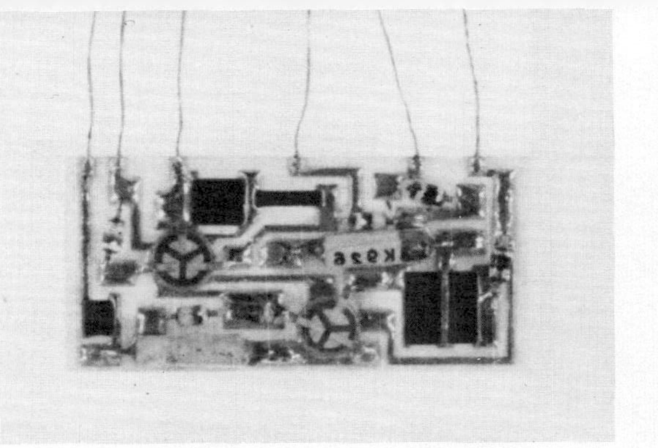

Figure 9.15 Relative ceramic printed circuit sizes under 1X, 4X, 10X and 20X magnification levels.

Table 9.1 Visual Field for Each Magnification Condition

Magnification Condition	Field of View	Depth of Field
Unaided vision	Complete CPC	Complete CPC
2X	Complete CPC	Complete CPC
4X	Complete CPC	Complete CPC
10X	0.79 inch	0.25 inch
20X	0.39 inch	0.10 inch

vision but the increase from 4X to 10X was both statistically and practically significant. On the basis of these findings, it was recommended that 10X magnification be provided for inspecting all quality characteristics of ceramic printed circuit assemblies.

Magnification Requirements of Labels

In the assembly and inspection of small devices, it is usually necessary to identify by number the components or elements that go to make up the device. In establishing a system for component identification, it is useful to know the level of magnification required to read labels of various sizes; for example, it would be most efficient to provide labels that can be read with the same level of magnification required for assembly or inspection operations.

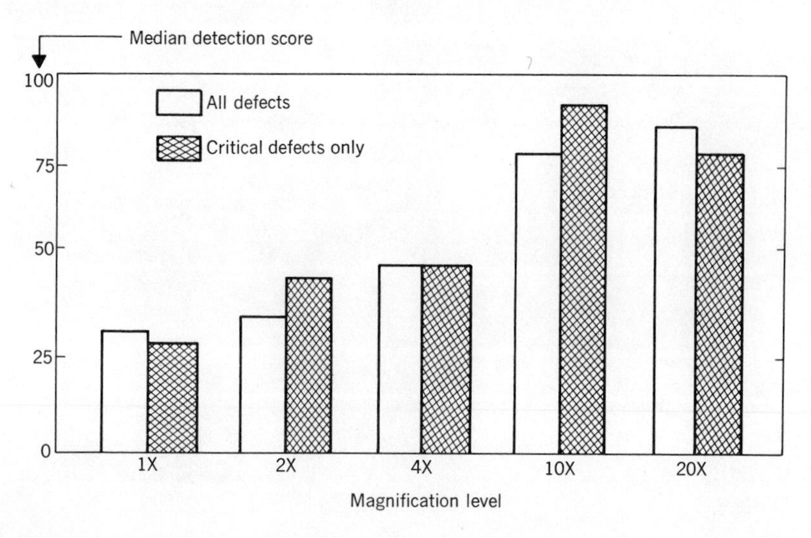

Figure 9.16 Comparison for median detection scores on all defects and on critical defects only.

Assuming that optimal numeral and letter styles are used, the magnification required for comfortably reading numbers and letters of any given size can be calculated. Research has indicated the optimal configurations and sizes for letters and numbers required for specified viewing distances [9]. Black characters of the following configuration should be placed on a white background: the width of the figure should be 80 percent of the height except for the number 1 where the width may be 20 percent of the height, stroke width should be 16 percent of the figure height, and spacing should be 16 percent of figure height. Since the minimum width of a number or letter for comfortable reading is 0.040 inch at a 20-inch viewing distance, the magnification required for comfortably reading numbers of any size can be calculated. The theoretical magnification (m) required to read a figure of width (w) can be determined from the equation $m = 0.040/w$.

Within this theoretical framework, a practical guide was developed for use in providing component identification. The magnification level and character sizes required for component labels of various lengths are shown in Table 9.2. Since the magnification instrument may cause certain reductions in legibility such as optical distortions of the image, decreases in brightness, and decreases in the contrast ratio, these theoretical calcu-

Table 9.2 Magnification Level and Character Size Required for Components of Various Lengths

Component Label Length*	Recommended Magnification	Character		
		Width*	Height*	Spacing*
0.025	20X	0.002	0.003	0.0005
0.050	10X	0.004	0.005	0.0008
0.075	6.5X	0.006	0.007	0.0011
0.100	5X	0.008	0.010	0.0016
0.125	4.5X	0.009	0.011	0.0018
0.150	3.5X	0.011	0.014	0.0022
0.175	3X	0.013	0.016	0.0025
0.200	2.5X	0.015	0.019	0.0030
0.225	2.3X	0.017	0.021	0.0034
0.250	2.1X	0.019	0.024	0.0039
0.300	1.7X	0.023	0.029	0.0046
0.350	1.5X	0.027	0.034	0.0055
0.400	1.2X	0.031	0.040	0.0064
0.450	Eye	0.035	0.044	0.0070
0.500	Eye	0.039	0.049	0.0078

* In inches.

lations were verified with some empirical evidence. Inspectors viewed characters throughout the range of dimensions specified in Table 9.2 and set their own magnification level at the point at which viewing was most comfortable. These data indicated that the theoretical values were not degraded by the magnification system. Therefore the data provided in Table 9.2 should be useful in establishing the size and configuration of component identification labels.

PRECISION MEASUREMENT AIDS

Precision measurement tasks such as those required for the inspection of machined parts involve skills different from those employed in visual inspection tasks. These skills include the operation of measurement devices such as gauges and micrometers, the use of mathematics, the interpretation of blue prints and the application of geometric concepts as well as certain visual inspection methods. On the surface it would appear that inspections involving precision measurement are much more objective in nature. This may be true but it does not seem to be a guarantee of high levels of inspection effectiveness.

Accuracy of Precision Measurement

In an extensive study of the accuracy of measurements made with certain precision instruments Lawshe and Tiffin [10] found that inspection accuracy was unexpectedly low. In this study a room consisting of 20 numbered booths was prepared; a different precision measuring instrument was placed in each booth together with a simplified working drawing that indicated the dimension to be measured with the instrument provided. An inspector used only those instruments that he used on his particular job. Each inspector made five measurements and then recorded his best judgment with respect to the dimension. His reading was then compared with actual dimensions that had been determined by means of ultraprecision instruments. Analyses of the data obtained indicated that in most cases less than 50 percent of the inspectors met the required standards of inspection accuracy. These results are shown in Table 9.3.

Visual Aids for Precision Measurement

In a recent study of the accuracy of machined parts inspectors, relatively low levels of inspection accuracy were again found [11]. In an initial part of this study, average defect detection performance was found to be about 40 percent with a range of individual detection levels from 25 to 80 percent. To identify the most promising areas of potential im-

Table 9.3 Percentages of Inspectors Who Met Established Standards of Accuracy in Making Precision Measurements

Instrument	Standard of Accuracy	Number of Inspectors	Percentage Meeting Standard
One-inch vernier micrometer	±0.0001	162	43
Two-inch vernier micrometer	±0.0001	138	17
Six-inch vernier micrometer	±0.0001	131	11
Three-inch (regular) micrometer	±0.001	146	64
Depth micrometer	±0.001	142	53
Inside micrometer	±0.001	113	66
Inside caliper and two-inch micrometer	±0.001	127	46
Inside caliper and six-inch micrometer	±0.002	112	9
Outside caliper and six-inch rule	±1/64	117	49
Vernier caliper (inside)	±0.001	113	42
Vernier caliper (outside)	±0.001	117	51

Adapted from C. H. Lawshe and J. Tiffin, The Accuracy of Precision Instrument Measurement in Industrial Inspection, *loc. cit.*

provement in precision measurement effectiveness, job sample performance data were obtained and analyzed to determine the relative frequency of various types of inspection errors. Primary areas requiring improvement were found to be detection of mislocated holes, thread gauging, and measurement of parallelism and concentricity. In reviewing these data and in observing the job of machined parts inspectors, difficulty was noted in going from the relatively complex blueprints available for the parts being inspected to the sequence of measurement operations required in inspecting these parts. It appeared necessary to bridge the gap between these complex drawings and the particular actions that a machined parts inspector was required to take. As a consequence of these observations, a study was initiated to determine the impact of visual aids in the form of simplified drawings on machined parts inspection performance and to compare the effect of visual aids with that of on-the-job training in these deficient areas.

Visual aids consisting of a series of simplified drawings were prepared for two sample parts which were the job sample measures used in the initial study of machined parts inspection effectiveness—a synchro bracket and a motor support. These parts are shown in Figure 9.17. Six visual aids were prepared for the bracket and eight for the support. The dimensions, tolerances, and inspection criteria for each characteristic were placed on the appropriate drawings to minimize the need for calculation and reference to other materials. The visual aids were also designed to reduce the probability of improperly setting up equipment or overlooking critical characteristics; this was accomplished by grouping similar items on a page and omitting any dimensions of characteristics that were not to be inspected. The visual aids were prepared by an experienced inspector who had not participated in the previous study; this eliminated the possibility of including information that would be specific to any of the defects contained in the job sample parts. An example of one of the visual aids prepared for the bracket is shown in Figure 9.18. A machined parts inspector is shown in Figure 9.19 using the blueprint to assist him in his inspection. He is also shown using one of the visual aids prepared for use in the same inspection task.

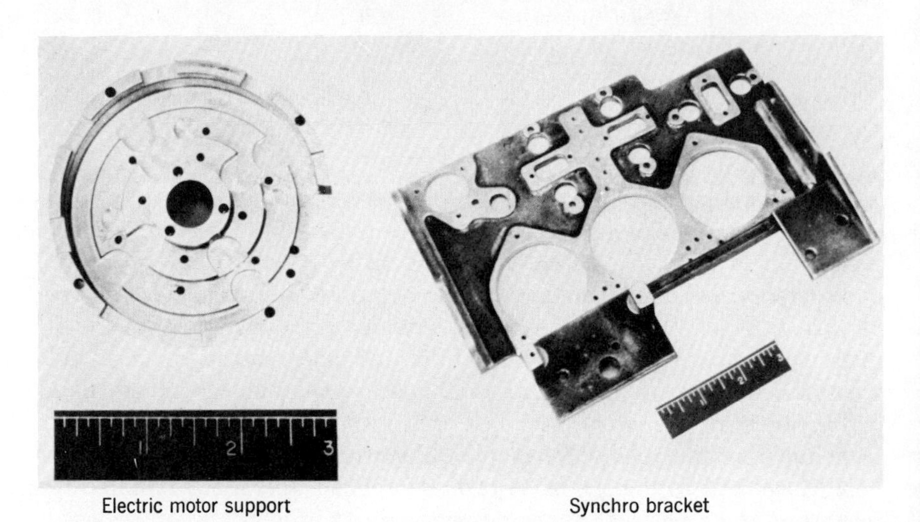

Figure 9.17 Job sample parts.

To evaluate the effectiveness of the visual aids, 26 inspectors were divided into four groups. The groups were matched in measures of inspection performance which had been obtained six months earlier. The first group served as a control for possible changes in accuracy that were not a direct result of the methods being tested; their performance was measured but they were given neither training nor the visual aids. The second group was given on-the-job training by the supervisor and the third group was given visual aids. A fourth group was given both training and visual aids. The inspection performance data obtained from the four groups were then analyzed to determine the relative effectiveness of the procedures. The results clearly indicated that the use of visual aids improved performance, that the use of training and visual aids improved performance significantly more than use of either approach alone. The combined effectiveness of the two experimental procedures was most evident in the detection of mislocated holes; a 450 percent increase was obtained. The combined use of training and visual aids also resulted in 100 percent detection of threaded hole defects and substantial improvements in the accuracy of parallelism and concentricity measurements. The overall results of the study are summarized in Figure 9.20.

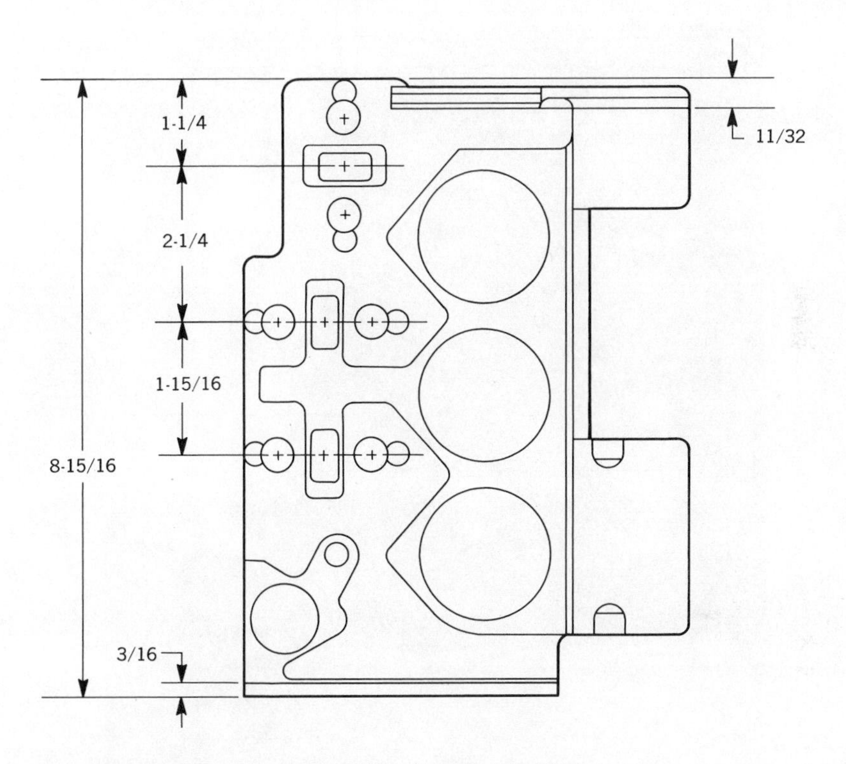

Figure 9.18 Sample visual aid.

It was concluded that high levels of precision measurement accuracy are likely to require both training and visual aids. The training content and visual aid format should be based on specific needs which have been identified by job sample performance measures. Visual aids should be designed to reduce inspection criteria to their simplest form and to assist the user in systematically inspecting for one type of defect at a time.

INFORMATION TRANSMITTAL FORMS

The information required for adequate management and control of complex industrial systems is obtained only at a significant cost. An appreciation of this cost is realized by considering the large numbers of employees involved in collecting, processing, and reporting information, together with the amounts of time spent by others in completing the forms and records associated with their jobs. As discussed earlier, the transmittal of information is a significant part of the inspector's job. Consequently

Figure 9.19 Machined parts inspector using blueprint (*a*) and visual aids (*b*).

there is need for a continuing evaluation of new information handling techniques as they become available. During the last several years many advances have been made in data processing techniques; computer access times have been significantly reduced, printing speeds have been greatly increased, and programming systems have become easier and cheaper to use. Improvements in recording and entering data into computer systems have, however, lagged behind advances in these other areas, although there has been some recent attention given to more direct data entry methods. Optical mark page readers, for example, have been developed; these devices are capable of transferring information contained on record

(b)

forms directly onto punched cards or into a data processing system. This approach requires information to be recorded at the source by marking the appropriate number, letter, or other labeled alternatives on an 8½ x 11 specially prepared form. The positions of the mark on the sheet provide the basis for data conversion. An example of an alphanumeric form prepared for this type of data transmittal is provided in Figure 9.21. The traditional form used for transmitting the same information is shown in Figure 9.22.

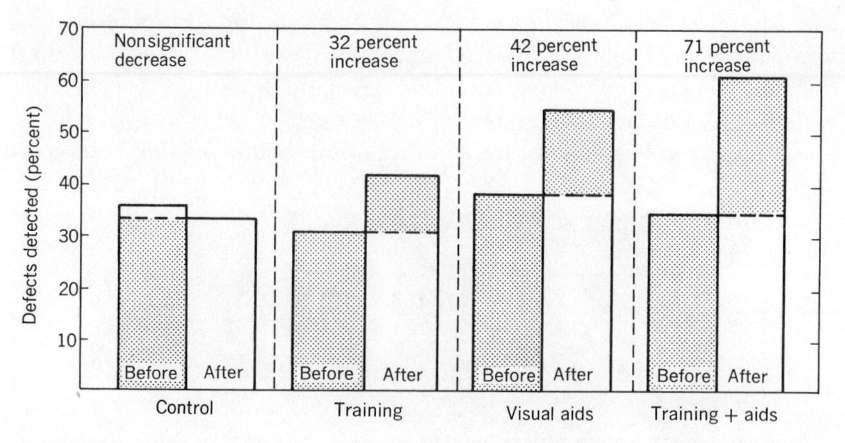

Figure 9.20 Improved machined parts inspection performance through the use of visual aids and training.

The two information transmittal forms described represent two different concepts in reporting inspection data. Therefore, in conjunction with the development of quality information systems, it is necessary to determine the relative speed and accuracy with which data can be transmitted by inspectors with each of the alternative forms. In a study directed toward such an evaluation, inspectors made information transmittals using both the alphanumeric form and the traditional form [12]. The source documents were typical production record cards. The study was designed so that any significant differences in performance would be a result of the type of form used. Before using either form each inspector was instructed in use of the form and was required to make five practice transmittals. Any errors made during the practice sessions were reviewed with the inspector. Measures of speed and accuracy were obtained for each inspector. Speed was measured by recording the number of minutes required to complete each set of 10 forms. Accuracy was measured by counting the number of errors made on each form.

Information transmittals with the alphanumeric form required twice as much time, on the average, as transmittals made with the traditional form. It took an average of 2.2 minutes for an alphanumeric transmittal and an average of 1.1 minutes for a traditional transmittal. There was no significant difference between forms in accuracy of transmittals. Initially, there were more errors made using the alphanumeric form; however, after practice, errors made on the two forms were numerically equal. It was concluded that use of the alphanumeric form would not result in any more transmittal errors than use of the traditional form, but that the alphanumeric form would take about twice as much time to complete.

QUALITY CONTROL INFORMATION TRANSMITTAL

QCIT NMBR.						DEPT.			WORK CENTER				INSP. OPER.			DAY		INSP. STAMP				MFG. LEAD			MECHANIC						JOB CONTROL NO.						
0	0	0	0	0	0	0	0	0	0	0	0		0	0	0		0	0	0	0	0	0	0	0	0	0	0	0	0	0	0	0	0	0	0	0	0
1	1	1	1	1	1	1	1	1	1	1	1	A	1	1	1	1	1	1	1	1	1	1	1	1	1	1	1	1	1	1	1	1	1	1	1	1	1
2	2	2	2	2	2	2	2	2	2	2	2	B	2	2	2	2	2	2	2	2	2	2	2	2	2	2	2	2	2	2	2	2	2	2	2	2	2
3	3	3	3	3	3	3	3	3	3	3	3	C	3	3	3	3	3	3	3	3	3	3	3	3	3	3	3	3	3	3	3	3	3	3	3	3	3
4	4	4	4	4	4	4	4	4	4	4	4	D	4	4	4		4	4	4	4	4	4	4	4	4	4	4	4	4	4	4	4	4	4	4	4	4
1	2	3	4	5	6																																
5	5	5	5	5	5	5	5	5	5	5	5	E	5	5	5		5	5	5	5	5	5	5	5	5	5	5	5	5	5	5	5	5	5	5	5	5
6	6	6	6	6	6	6	6	6	6	6	6	F	6	6	6		6	6	6	6	6	6	6	6	6	6	6	6	6	6	6	6	6	6	6	6	6
7	7	7	7	7	7	7	7	7	7	7	7	G	7	7	7		7	7	7	7	7	7	7	7	7	7	7	7	7	7	7	7	7	7	7	7	7
8	8	8	8	8	8	8	8	8	8	8	8	H	8	8	8		8	8	8	8	8	8	8	8	8	8	8	8	8	8	8	8	8	8	8	8	8
9	9	9	9	9	9	9	9	9	9	9	9	I	9	9	9		9	9	9	9	9	9	9	9	9	9	9	9	9	9	9	9	9	9	9	9	9

INSP. OPER.: FINAL

PART NUMBER											SERIAL PREFIX						NUMBER				GO NUMBER				
0	0	0	0	0	0	0	0	0	0	0	0	MARK IF S-Z	0	MARK IF S-Z	0	MARK IF S-Z	0	0	0	0	0	0	0	0	0
1	1	1	1	1	1	1	1	1	1	1	1	A J	1	A J	1	A J	1	1	1	1	1	1	1	1	1
2	2	2	2	2	2	2	2	2	2	2	2	B K S	2	B K S	2	B K S	2	2	2	2	2	2	2	2	2
3	3	3	3	3	3	3	3	3	3	3	3	C L T	3	C L T	3	C L T	3	3	3	3	3	3	3	3	3
4	4	4	4	4	4	4	4	4	4	4	4	D M U	4	D M U	4	D M U	4	4	4	4	4	4	4	4	4
5	5	5	5	5	5	5	5	5	5	5	5	E N V	5	E N V	5	E N V	5	5	5	5	5	5	5	5	5
6	6	6	6	6	6	6	6	6	6	6	6	F O W	6	F O W	6	F O W	6	6	6	6	6	6	6	6	6
7	7	7	7	7	7	7	7	7	7	7	7	G P X	7	G P X	7	G P X	7	7	7	7	7	7	7	7	7
8	8	8	8	8	8	8	8	8	8	8	8	H Q Y	8	H Q Y	8	H Q Y	8	8	8	8	8	8	8	8	8
9	9	9	9	9	9	9	9	9	9	9	9	I R Z	9	I R Z	9	I R Z	9	9	9	9	9	9	9	9	9

WORK AUTH: PROD, ESWA, M&R

TYPE INSP: NEW, RESUB, AUDIT

PLAN: MIL-STD., 100%, SPECIAL, MAT-REV

LOT SIZE				SAMPLE SIZE				LOT/SAMPLE DEFECTIVES			
0	0	0	0	0	0	0	0	0	0	0	0
1	1	1	1	1	1	1	1	1	1	1	1
2	2	2	2	2	2	2	2	2	2	2	2
3	3	3	3	3	3	3	3	3	3	3	3
4	4	4	4	4	4	4	4	4	4	4	4
5	5	5	5	5	5	5	5	5	5	5	5
6	6	6	6	6	6	6	6	6	6	6	6
7	7	7	7	7	7	7	7	7	7	7	7
8	8	8	8	8	8	8	8	8	8	8	8
9	9	9	9	9	9	9	9	9	9	9	9

Figure 9.21 Alphanumeric form.

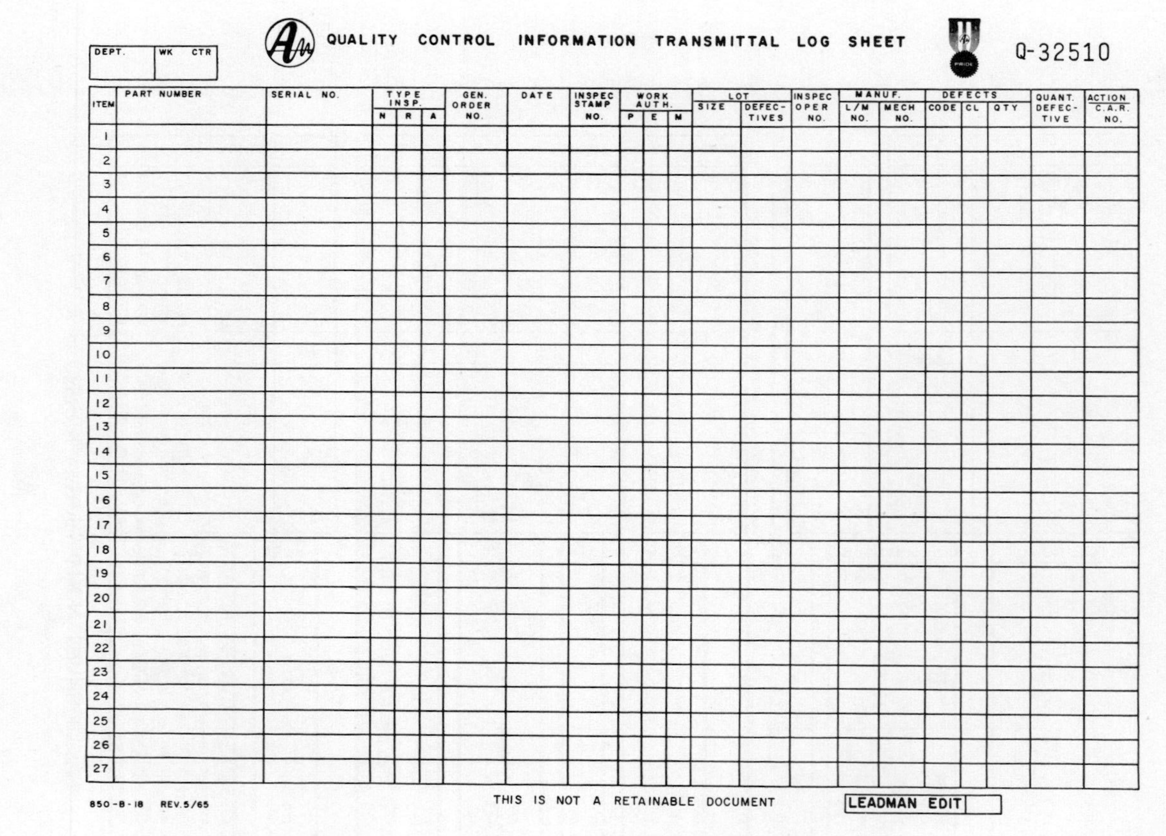

DEPT. | WK CTR

QUALITY CONTROL INFORMATION TRANSMITTAL LOG SHEET

Q-32510

ITEM	PART NUMBER	SERIAL NO.	TYPE INSP.			GEN. ORDER NO.	DATE	INSPEC STAMP NO.	WORK AUTH.			LOT		INSPEC OPER NO.	MANUF.		DEFECTS			QUANT. DEFEC-TIVE	ACTION C.A.R. NO.
			N	R	A				P	E	M	SIZE	DEFEC-TIVES		L/M NO.	MECH NO.	CODE	CL	QTY		
1																					
2																					
3																					
4																					
5																					
6																					
7																					
8																					
9																					
10																					
11																					
12																					
13																					
14																					
15																					
16																					
17																					
18																					
19																					
20																					
21																					
22																					
23																					
24																					
25																					
26																					
27																					

850-B-18 REV.5/65

THIS IS NOT A RETAINABLE DOCUMENT

LEADMAN EDIT

Figure 9.22 Traditional form.

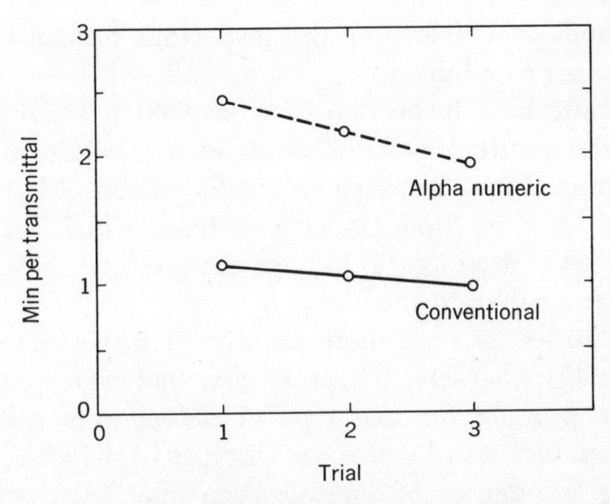

Figure 9.23 Average transmittal times for alphanumeric and traditional forms.

These results are shown in Figures 9.23 and 9.24. The practical significance of the additional time required to complete alphanumeric transmittals must be determined in a cost effectiveness analysis of the specific application. Cost reductions obtained from replacing keypunch equipment and personnel with optical mark page readers may more than offset the cost increases resulting from the additional transmittal time required by inspectors. The results of the experimental study provided the critical performance data needed for such an analysis.

SUMMARY

1. New concepts in inspection tools and in methods employed by the inspector in his job have demonstrated a significant impact on inspection performance. The basic idea is to improve inspection performance through

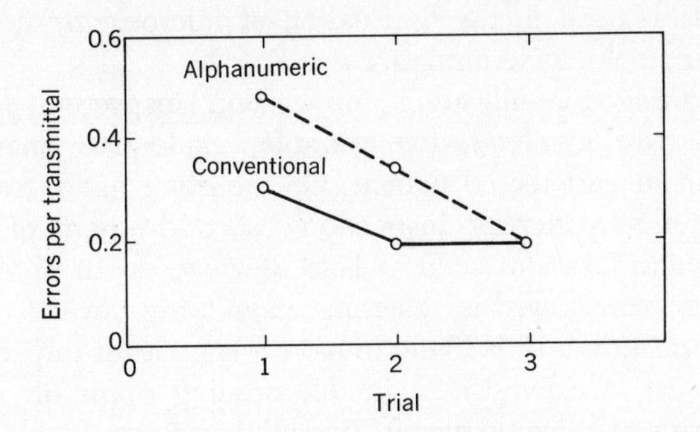

Figure 9.24 Average errors per transmittal for alphanumeric and traditional forms.

tools and methods that overcome the inspector's human limitations and enhance his human capabilities.

2. In a scanning type inspection task, inspection accuracy is likely to be greater if the product is scanned while it is stationary rather than while it is moving. If it is necessary to conduct a scanning type inspection of a moving product or products such as those which are passing on a conveyor belt, the best method is to have the products move laterally past the inspector rather than toward him.

3. Scanning inspections are more effective if inspectors attend to only one type of quality characteristic at a time; that is, an inspector should scan the entire product for one type of defect at a time rather than searching the product area by area for all types of defects.

4. Increasing the degree of specialization may be a necessary means of gaining the most benefit from higher levels of magnification. Increased specialization is related to the characteristic-by-characteristic scanning method in that the inspector's attention at any time is focused on just a few possible defects. It has been found that increasing magnification beyond a certain point without changing inspection procedures to accommodate for the new conditions may actually reduce inspection accuracy rather than increase it.

5. For inspection tasks that require the detection of defects in a large array of similar items, such as the electrical circuitry on circuit boards, overlay tools can be designed and used to cause defects to "pop out" at the inspector. The use of overlay tools has resulted in significant improvements in inspection performance.

6. Because of the special capability of the human being to integrate what he sees, effectiveness can be increased by structuring the inspection task so that the inspector makes direct comparisons between quality characteristics and quality standards presented to him simultaneously. Comparison techniques have been employed successfully in the inspection of materials used in the fabrication of microelectronic devices and for making fine color discriminations.

7. In providing magnification for certain inspection tasks, several considerations are involved; for example, inadequate magnification is likely to result in undetected defects. On the other hand, excessive levels of magnification may increase inspection costs and reports of false defects. In addition, considerations such as field of view, depth of view, and distortion must be considered in selecting magnifying devices.

8. If an inspection task is likely to have a significant impact on product quality and cost, an investment in determining optimum magnification levels may result in a sizable payoff. Since the optimum level depends upon

a number of factors, the recommended approach is to obtain measures of inspection performance under alternative levels of magnification.

9. Investigations of the accuracy of performing precision measurement inspection tasks have indicated that inspection accuracy is typically much lower than expected. In one study less than 50 percent of the inspectors met the required standards of inspection accuracy. In another study average defect detection performance was found to be only about 40 percent.

10. Both on-the-job training and visual aids may be employed to significantly increase the accuracy of measurement type inspections. Training content should be based on specific needs which have been identified by job sample performance measures. Visual aids should be designed to reduce inspection criteria to their simplest form and to assist the user in systematically inspecting for one type of defect at a time.

11. The transmittal of quality information is a significant part of the inspector's job. Consequently there is need for a continuing evaluation of new information handling techniques as they become available. One approach to information transmittal is the use of alphanumeric forms which require information to be recorded by marking the appropriate number, letter, or other labeled alternative. Because the positions of the marks on the sheet provide the basis for data conversion, the information can be fed directly into the computer by means of mark page readers. Although this approach eliminates several steps in the data transmittal process, information recording is slower and no more accurate than with traditional forms.

REFERENCES

[1] Douglas H. Harris, Effect of Equipment Complexity on Inspection Performance. *J. appl. Psychol.*, 1966, **50**, 236–237.

[2] J. Tiffin and H. B. Rogers, The Selection and Training of Inspectors. *Personnel*, 1941, **18**, 14–31.

[3] N. C. Kephart and G. G. Besnard, Visual Differentiation of Moving Objects. *J. appl. Psychol.*, 1950, **34**, 50–53.

[4] Frederick B. Chaney and Douglas H. Harris, Human Factors Techniques for Quality Improvement, *Ann. ASQC Tech. Conf.* 1966, **20**, 400–413.

[5] Ernest E. Sadler, The Effect of an Overlay Field on Visual Inspection Judgments. A paper presented to the Western Psychological Association, April 28–30, 1966.

[6] J. P. Guilford, *Psychometric Methods*, 2nd Ed., New York: McGraw-Hill, 1954.

[7] Robert M. Springer and Douglas H. Harris, Human Factors in the Production of Microelectronic Devices. A paper presented at the Eighth Annual IEEE Symposium on Human Factors in Electronics. Palo Alto, California, May 4, 1967.

[8] J. R. Simon, Magnification as a Variable in Subminiature Work. *J. appl. Psychol.*, 1964, **48**, 22–24.

[9] C. T. Morgan, *Human Engineering Guide to Equipment Design.* New York: McGraw-Hill, 1963. Pp. 102–103.

[10] C. H. Lawshe and J. Tiffin, The Accuracy of Precision Measurement in Industrial Inspection. *J. appl. Psychol.*, 1945, **29**, 413–419.

[11] Frederick B. Chaney and Kenneth S. Teel, Improving Human Performance Through Training and Visual Aids. *J. appl. Psychol.*, 1967, **51**, 311–315.

[12] Douglas H. Harris, Information Transmittal Performance with Alphanumeric Forms. *J. Eng. Psychol.*, 1965, **4**, 86–91.

X
INSPECTION DECISION AIDS

Decision making is the focal point of inspection work. At this point a quality characteristic has been compared with the quality standard in preparation for a judgment of whether or not the characteristic conforms to the standard. Beyond this point action is taken in accordance with the decision. There are two basic types of decision-making errors that an inspector can make. He can fail to report a true defect or he can report something that is not a defect. A true defect is not reported because the defect is not observed or, if it is observed, because an error in judgment is made. A false detection, on the other hand, results solely from an error in judgment.

A problem in inspection decision making is that of providing inspectors with meaningful and usable quality standards for their judgments of conformity. Most often inspectors must refer to one or more of a large number of process and product specifications for information to use in decision making. In doing so they typically must read and understand the specification and, provided the quality standards are sufficiently well described, transform what they read into some framework or mental image which defines the borderline between acceptable and unacceptable. Because of the number of characteristics involved, the difficulty of translating written words into useful images, and human limitations in memory and attention, it is not easy to provide useful quality standards for the inspector.

The size of the inspection decision-making problem was illustrated in the previously cited series of studies [1]. These studies included quantitative data with which to explore the problem of inspector decision making. The average percentages of unreported defects and of false detections were obtained from inspections of each of 10 different items of electronic equipment. Each equipment sample contained a representative set of defects. Each item of equipment was inspected by eight or more inspectors experienced in its inspection and regularly assigned to inspecting items of this type. Table 10.1 lists the average percentages of unreported defects and false detections for each item. As seen from the table, these results indicate inspection decision making to be a significant problem in inspection effectiveness. About 50 percent of the legitimate defects in these items went unreported and, of those that were reported, more than 40 percent were false detections. These results indicated the need to clarify the borderline between acceptable and unacceptable quality characteristics.

Table 10.1 Percentages of Defects Not Detected and False Detections in Inspections of Various Equipment Items

Equipment Item	Number of Inspectors	Percent of Defects Not Detected	Percent of False Detections
Autonavigator chassis	12	86	50
Autonavigator module	19	73	43
Alignment set	8	71	70
Radar module	8	61	47
Microelectronic device	10	51	15
Printed circuit assembly	10	46	23
Wire harness	8	40	68
Circuit board assembly	8	31	38
Ceramic circuit chip	10	29	19
Etched circuit board	13	26	25

In addition to the need for initially clarifying the differences between acceptable and unacceptable characteristics, there is the need for maintaining an unbiased "perceptual set" in the inspector. An inspector's judgment may be influenced by factors in his surroundings that are not related to the established quality standards; for example, an inspector's interactions with other people in inspection, manufacturing, and engineering operations may cause his judgments of conformance to differ from those dictated by the quality standards. The potential impact of these social and organizational factors on the accuracy of inspection decision making has been demonstrated in a number of studies[2,3,4]. As a result, the need for an inspector to maintain an unbiased frame of reference in the face of these potentially biasing factors also underlines the need for meaningful and usable quality standards.

THE USE OF LIMIT SAMPLES

As discussed and illustrated in Chapter IX, human judgments are usually most accurate when they involve direct comparisons; for example, it is easier by means of a direct comparison to determine that one object is heavier or smoother than another than to determine, without a comparison, that an object is too heavy or not smooth enough. Therefore one way of improving the accuracy of inspection decisions would be to provide an example against which the characteristic in question could be compared. The example selected could be just barely acceptable with respect to a given quality characteristic. A product could be compared with the example, or standard, and if it were less acceptable than the standard it would be rejected. An example used in this manner might be called a limit sample because it specifies the limit of product acceptability with respect to a given quality characteristic.

The effectiveness of using limit samples was illustrated in a study of the inspection of glass panels for the fronts of television picture tubes [5]. Inspectors had been accepting or rejecting these panels in terms of their mental image of acceptable versus unacceptable panels. After establishing an inspection procedure which employed the use of limit samples, inspection accuracy increased 76 percent. In this procedure several limit samples were employed by an inspector, one for each of several types of defects. The limit samples were actual glass panels which had been systematically selected as the limit of acceptability for each type of defect.

The major drawback in the use of limit samples is a practical one. How do you have the required example readily available when needed for a direct comparison? If there are many quality characteristics involved, the mechanics of providing limit samples for each type of defect are burdensome and time consuming. As a consequence alternative approaches which utilize the limit sample concept but which overcome these practical difficulties may offer a more useful solution.

PHOTOGRAPHIC DECISION AIDS

The use of photographs as limit samples has been investigated as a method of overcoming some of the display and handling problems of using actual products. In a study of the value of photographic decision aids photographic limit samples were developed for use in inspecting solder connections [6]. The basic idea behind the development of photographic aids for this purpose was that solder connections exist on an acceptable/unacceptable continuum with respect to each of several quality standards. Items at the edges of the continuum are clearly unacceptable or clearly acceptable, and, consequently, provide no decision problems. The problem area is the middle range where quality characteristics are close to the borderline. The following approach was taken in developing photographic aids:

1. The degree of acceptability of each of a sample of quality characteristics was evaluated and an acceptability value assigned to each on the basis of the evaluation. An acceptability value for each characteristic was derived from ratings of each by four senior inspectors. Each inspector gave each characteristic a rating of 1, 2, 3, or 4, in accordance with the following scale.

4 Acceptable
3 Borderline—acceptable
2 Borderline—unacceptable
1 Unacceptable

The acceptability value for a given characteristic was obtained by computing the average rating given that characteristic by the four inspectors.

2. Certain representative characteristics with values near either side of the acceptable/unacceptable borderline were selected and photographed.

3. Pictures of borderline-acceptable characteristics and borderline-unacceptable characteristics were taken. These were appropriately labeled and used as aids for inspection decisions.

To evaluate the effectiveness of photographic aids developed in this manner, acceptability values were obtained for each of a sample of 116 solder connections. Solder connections had been shown repeatedly in previous studies to be a major decision-making problem area. Consequently a study of the impact of photographic aids on inspections of solder connections was considered to be a good test of the usefulness of the aids. On the basis of the acceptability values obtained, eight solder connections were selected to be photographed; four of these were in the borderline-acceptable category, and four were in the borderline-unacceptable category. Eight- by 10-inch photographs were prepared. A sample of the resulting photographs is shown in Figure 10.1.

Three groups inspected a sample of 54 solder connections without use of the photographic inspection aids and then inspected a different sample of 54 using the inspection aids. The experimental design eliminated any effects which might be introduced because of the differences between the two samples of solder connections or differences in the order in which the two samples were inspected. Each group consisted of six inspectors experienced in inspecting solder connections of the type provided.

Significantly greater agreement among inspectors was obtained when photographic inspection aids were used than when they were not. The amount of agreement was computed by determining the percentage of cases in which a set of three inspectors agreed on the acceptability of a characteristic. The total number of cases was 324 (six groups of three inspectors times 54 characteristics) for each condition. In addition, the

Figure 10.1 Sample photograph aid showing minimum acceptable solder contamination.

amount of agreement at various levels of solder acceptability was computed. As shown in Figure 10.2, inspection agreement was increased most in the borderline area where agreement was the lowest initially. In addition, a significant increase in the percentage of correct decisions was found with the use of the photographic aids.

PHOTOGRAPHIC QUALITIES

Results of the initial study were encouraging because they indicated that use of systematically prepared photographs can significantly increase agreement among inspectors and can help inspectors identify acceptable and unacceptable quality characteristics. It was also encouraging in that the greatest increase in agreement involved characteristics close to the borderline. Since the initial study did not systematically investigate the effect of photograph qualities on inspection decisions, a second study was undertaken [7]. The objective of this study was to determine the effect of the following photograph qualities on inspection decision making: (a) color versus black and white, (b) size—4 x 5 inch photograph versus 8 x 10 inch, (c) distance of examples from the acceptable/unacceptable borderline, and (d) combinations of the above.

Eight sets of photographs were developed, one set for each combina-

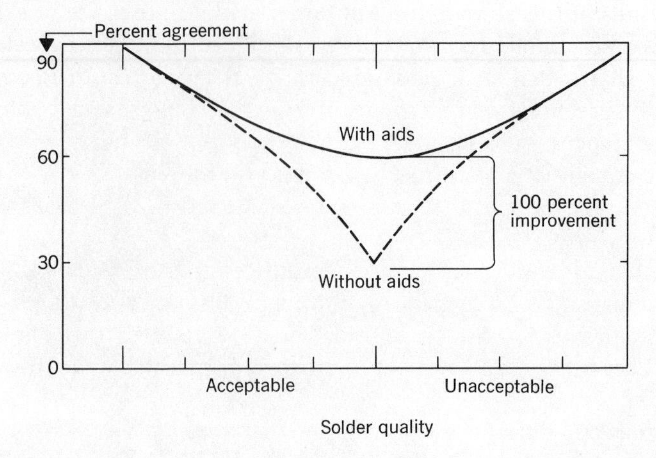

Figure 10.2 Increase in agreement among inspectors resulting from use of photographic aids.

tion of photograph qualities. As a basis for developing the photographs, acceptability values were computed from evaluations of more than 1300 solder connections by each member of a panel of five experts. Within each set there were two photographs for each solder characteristic—one showing a borderline-acceptable condition and one showing a borderline-unacceptable condition. A sample of 24 inspectors looked at solder connections with and without the photograph aids. Performance under each of the nine conditions—use of no photographs plus use of each of the eight photograph sets—was evaluated in terms of different measures of decision-making performance; they are described as follows:

1. *Decision errors.* Two types of decision errors are possible—accepting bad items and rejecting good ones. Performance was measured by counting the number of errors of both types.

2. *Decision agreement.* Performance was measured by determining the degree to which different inspectors made the same decisions. The 24 participants in the study were considered, for purposes of this analysis, as eight groups of three inspectors each. Agreement occurred when the three inspectors in a group all accepted or all rejected a particular solder connection.

3. *Diagnostic errors.* When a solder connection was indicated to be unacceptable, the reason was also indicated by recording the defect category—excessive, insufficient, cold, overheated, disturbed, poor flow, void, or foreign material.

4. *Diagnostic agreement.* Performance was measured by determining the degree to which different inspectors made the same diagnosis. Agreement occurred when a group of all three made the same diagnosis.

The results of this study gave additional evidence that use of systematically developed photographs results in significantly better inspection performance than when such photographs are not used. Both decision and diagnostic agreement were significantly increased by using photographs. These differences were both statistically and practically significant. The following conclusions were reached regarding the characteristics of color and size.

1. Black and white photographs produced performance as good as color photographs with respect to all four performance measures.
2. Photographs 4 x 5 inches in size were as effective as photographs 8 x 10 inches in size with respect to all four performance measures.

No significant differences were found between near and far examples in their effect on decision errors, diagnostic errors, or diagnostic agreement. A significant difference, however, between near and far examples was found in their effect on decision agreement. Near examples were found to be superior to far examples in defining the acceptable/unacceptable borderline. A properly defined borderline should result in a minimum amount of bias with respect to decision errors. One would expect as many errors to be made in rejecting good items as in accepting bad items, if the borderline were properly defined. Any significant deviation from a 50–50 split would indicate that inspectors' interpretations of the borderline location are different from that provided by the photographs and, therefore, that communication of the borderline loaction is inadequate. The no photograph conditions and use of photographs with far examples resulted in significant deviations from the optimal 50–50 split, whereas use of photographs with near examples did not. These results are illustrated in Figure 10.3.

HOW TO DEVELOP PHOTOGRAPHIC AIDS

As a result of the research conducted to develop and evaluate photographic inspection aids, specific guidelines can be made available for developing and using photographs to aid inspection decision making. A step-by-step approach that can be used to develop photographic aids for any set of quality characteristics is as follows:

1. *Establish a panel of experts.* The panel of experts should consist of about four persons, each knowledgeable about the characteristics for which the photographic aids are to be made. The panel members should represent potentially different points of view; a good panel might consist of a quality control engineer, an inspection supervisor, an audit inspector, and a staff member of the Quality Control Laboratory.

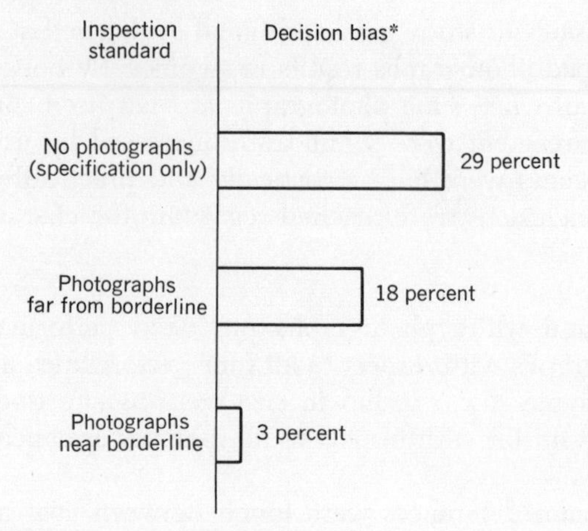

Figure 10.3 The value of photographic aids in reducing decision bias.

2. *Obtain a sample of items.* The sample should contain characteristics having a full range of acceptability, from clearly acceptable to clearly unacceptable. The bulk of the characteristics, however, should be relatively near the acceptable/unacceptable borderline. An attempt should also be made to include the various types of possible defects.

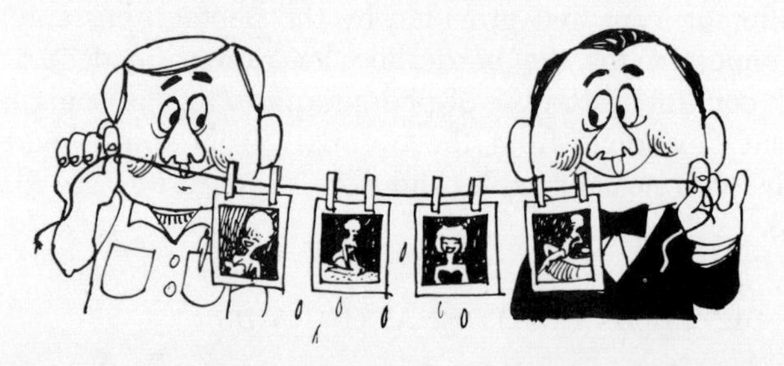

3. *Construct rating sheets.* A means of recording the rating given to each characteristic should be provided. A rating sheet consisting of an outline of the item and an indication of exactly where each characteristic is located should serve this purpose. An example of a rating sheet made up for a set of solder connections on a circuit board is shown in Figure 10.4. The locations of the characteristics (solder connections) are indicated by black dots. The circle attached to each dot provides a place for the rater to indicate his evaluation of that connection.

4. *Have each panel member rate each characteristic for acceptability.* The following scale may be used for obtaining acceptability ratings:

4 Acceptable
3 Acceptable—borderline
2 Unacceptable—borderline
1 Unacceptable

Each expert on the panel should examine each characteristic and assign it a value of 1, 2, 3, or 4 in accordance with the acceptability scale. Each rating should be performed individually; a rater should not know what ratings the other panel members have assigned. The end result of this step will be a set of rating sheets completed by each member of the panel.

5. *Compute an acceptability value for each characteristic.* The acceptability value for each characteristic is obtained by adding all the rating values given to the characteristic and dividing by the number of raters; for example, if a solder joint received ratings of 4, 3, 4, and 2 by four raters, the acceptability value would be computed by adding $4 + 3 + 4 + 2 = 13$, and dividing 13 by 4. The acceptability value thus obtained would be 3.25. The possible range of acceptability values is 1.00 to 4.00. The borderline between acceptable and unacceptable is 2.50.

6. *Select characteristics to be photographed.* To define the acceptable/unacceptable borderline for any characteristic, select two items having the required characteristic—an acceptable one having a value between 2.50 and 3.00 and an unacceptable one having a value between 2.00 and 2.50. Additional factors, such as representation of particular defect types, may

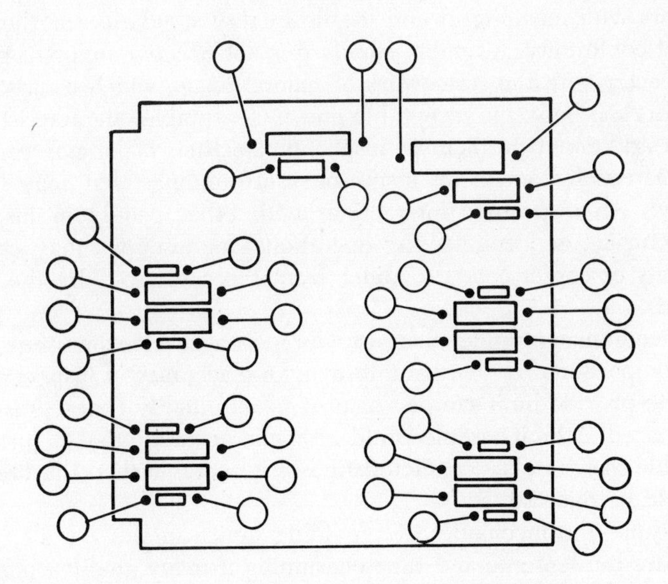

Figure 10.4 Quality characteristics rating sheet.

be considered in further narrowing the selection of characteristics. An alternative way of defining the borderline would be to use only the borderline-acceptable item with the instructions that anything less acceptable should be rejected.

7. *Photograph the characteristics.* To aid the photographer the angle at which the photograph is to be taken should be specified. One of the panel members should review each characteristic and should indicate, on a rating sheet, the angle at which each characteristic should be photographed.

8. *Compare the photographs and actual characteristics.* Using the resulting photographs instead of the actual characteristics, obtain acceptability values for each selected characteristic by repeating steps 4 and 5. Photographs which produce acceptability values outside the appropriate 2.50–3.00 or 2.00–2.50 range should be replaced. Photographs to replace the discarded ones should be obtained by photographing the same characteristics again or selecting and photographing comparable characteristics.

SUMMARY

1. An inspector can make two basic types of decision-making errors. He can fail to report a true defect or he can report something that is not actually a defect.

2. A major problem in inspection decision making is that of providing inspectors with meaningful and usable quality standards for their judgments of conformity. A quality standard is not effective unless it provides the inspector with the framework or mental image which clearly defines the borderline between acceptable and unacceptable characteristics.

3. An additional problem in inspection decision making is caused by the numerous factors in an inspector's surroundings that may bias his judgment. An inspector's interactions with other people in inspection, manufacturing, and engineering operations, for example, may cause his judgments of conformance to differ from those dictated by the quality standards.

4. Since human judgments are usually most accurate when they involve direct comparisons, one way of improving the accuracy of inspection decisions is to provide limit samples against which quality characteristics can be compared. A limit sample could be a characteristic that is just barely acceptable. Any quality characteristic less acceptable than the limit sample would be rejected.

5. Because the mechanics of providing limit samples for all types of defects are burdensome and time consuming if many quality characteristics are involved, the use of photographs as decision aids is probably a

more practical way of implementing the concept of the limit sample. The use of photographic aids has been shown to improve inspection decision making by as much as 100 percent for the more difficult borderline cases. In addition, the use of photographs to portray quality standards has been found to minimize the amount of bias present in inspection decisions.

REFERENCES

[1] Douglas H. Harris, Effect of Equipment Complexity on Inspection Performance, *J. appl. Psychol.*, 1966, **50**, 236–237.

[2] R. M. McKenzie, On the Accuracy of Inspectors. *Ergonomics,* 1958, **1**, 258–272.

[3] A. E. M. Seaborne, Social Effects on Standards in Gauging Tasks. *Ergonomics,* 1963, **6**, 205–209.

[4] K. F. H. Murrell, *Human Performance in Industry*. New York: Reinhold, 1965. Pp. 418–429.

[5] Martha L. Kelly, A Study of Industrial Inspection by the Method of Paired Comparison. *Psychol. Monogr.*, No. 394, 1955.

[6] Kenneth S. Teel and Douglas H. Harris, New Applications of Engineering Psychology. Paper presented at the Western Psychological Association Convention, Honolulu, June 1965.

[7] James L. Thresh and John S. Frerichs, Results Through Management Application of Human Factors. Paper presented at the American Society of Quality Control Technical Conference, New York, June 1966.

XI
INSPECTOR SELECTION

There are large differences among people in their ability to perform inspection tasks. Some inspectors are able to detect many more of the defects present in an item than others. This conclusion is based on the results of a number of studies of inspection performance. Some of these findings are illustrated in Figure 11.1 for each of several general types of electronic item. As shown, the average performance for some types of items is much better than the average performance for other types of items; however, in each case the performance of some inspectors is several times better than the performance of other inspectors.

Large individual differences are not necessarily restricted to tasks involving visual comparisons and subjective judgments; large differences among inspectors performing precision measurement tasks have been noted also. In a study of machined parts inspection performance the best inspector detected four times as many defects in a sample of machined parts as the poorest inspector [1]. The distribution of the percentages of defects identified by a sample of 26 experienced machined parts inspectors is shown in Figure 11.2.

These results indicate that better methods of selecting people for inspection tasks are likely to result in significant improvements in inspection performance. Because the inspectors who participated in these studies were comparable in training and experience, it would appear that the differences in performance essentially reflect basic aptitude differences for inspection work. These results therefore highlight the need for selecting people who have better than average aptitudes for inspection work.

SELECTION METHODS AND THEIR USE

The objective of personnel selection methods is to obtain and use information in a way that will maximize success on the job. There have been no personnel selection methods devised to date that will assure the correct decision every time; however, there has been success in develop-

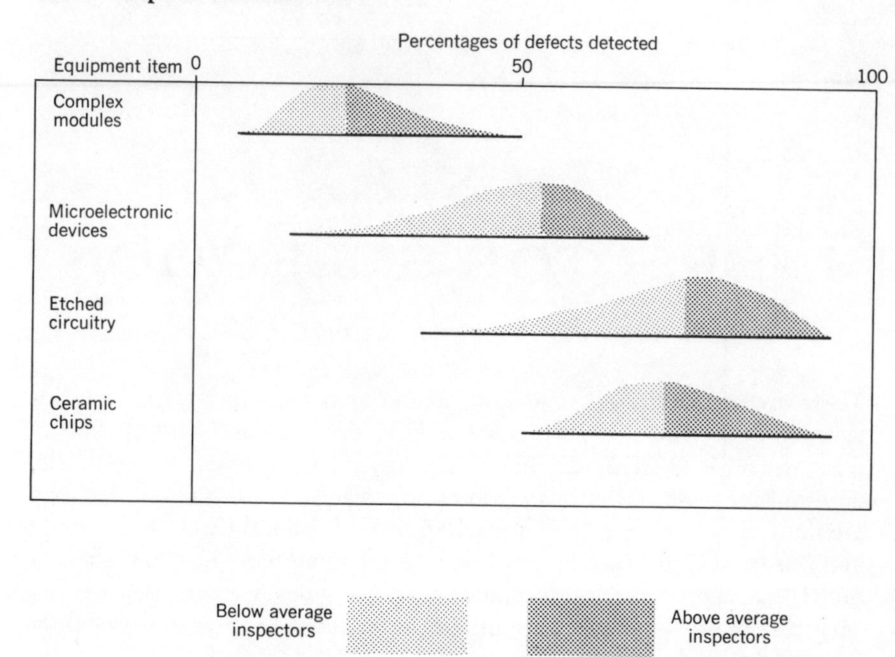

Figure 11.1 Distributions of inspection ability for different items.

ing and using selection methods to increase the "batting average" for making more accurate personnel selections. Some of the selection methods, such as interviews, background history information, reference checks, and personnel tests, have been used successfully for various jobs. These methods are essentially a means of obtaining information about someone about whom little is known. Obviously, if the person in question was someone whose performance and capabilities were well known, there would be little need for additional information. Some selection methods work better for certain types of job than for others. In the selection of salesmen the interview may be a very useful technique because it gets at the heart of such characteristics as appearance and oral communications ability which are important in sales work. On the other hand, the interview probably would not be very helpful for selecting inspectors because very little can be learned through the interview about how well a person will be able to detect defects. The use of personnel tests which measure the aptitudes required for successfully performing inspection tasks would probably be best.

A personnel test is a method used for obtaining a sample of behavior under controlled conditions. A test may take many forms. It may be a

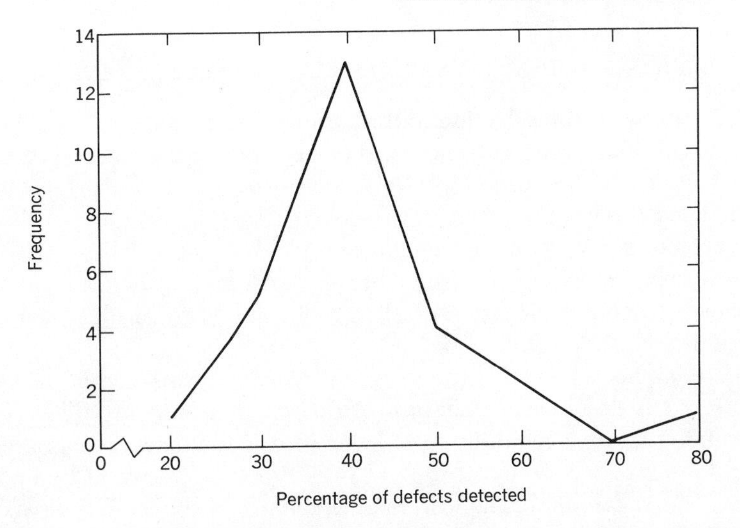

Figure 11.2 Distribution of percentages of defects detected by 26 machined parts inspectors.

series of questions which must be answered or it may be a task to be performed, such as a test of manual dexterity which involves picking pins out of a dish and putting them into holes. A test may even be in the form of a machine which makes certain demands upon the person being tested, such as a device used to obtain measures of visual acuity. In using the information provided by a test or series of tests for making personnel decisions, a person's test score is usually compared with the scores of others who have taken the test. If the person's score is in the range of scores of those people who have been successful on the job, the person would be assigned to the job.

Personnel tests are not necessarily useful even when they are valid. If, for example, an excellent job is being done in selecting people through the use of other methods, such as historical information or interviews, then there would be very little value in trading the present method for a test. Also, if there were many more jobs to fill than applicants for the jobs and the inclination to hire almost anyone prevailed, then a personnel test would probably not be very useful for selection purposes. Under these circumstances, however, a personnel test could still be very useful in matching the people who were hired with particular job specifications; for example, persons with inspection aptitude would be assigned to inspection jobs and people with other aptitudes to those jobs which most seemed to fit them.

VALIDATION OF PERSONNEL TESTS

To be useful in the selection and placement of personnel, a test has to be both valid and reliable. The validity of a personnel test refers to its built-in design to measure objectively what it says it is measuring. A test can be used safely only when it has demonstrated its validity. Although there are several ways to determine test validity, the general idea is to compare test scores with measures of job performance; if the test scores are shown to be significantly correlated with job performance measures, the test is considered to be valid.

The reliability of a test is the degree to which it consistently measures whatever it is designed to measure. Although there are several ways of determining the test reliability, the basic idea is the same. Scores are obtained for alternative forms of the test from the same people at two points in time. The reliability of the test, then, is the extent to which the two sets of scores are correlated.

It is often necessary to combine several tests into a test battery for use in selection of personnel for certain types of jobs, because there are several aptitudes required for success on the job. The idea in developing a test battery for a particular purpose is to include a set of tests which have significant validities for the job and low correlations with each other. The resulting battery of tests then is likely to correlate much higher with measures of job performance than would any of the tests individually.

Until recently very little attention has been given to the development of tests and test batteries specifically for the selection and placement of inspection personnel. Attempts to employ standard personnel tests, such as tests of mental alertness, have been generally unsuccessful. In the remainder of this section two successful efforts in developing tests specifically for certain inspection jobs are described. The first study describes the development and validation of a test for the selection of personnel for scanning inspection tasks. The second study describes the development and validation of a test battery for use in selecting people for measurement type inspection tasks.

SELECTION OF PERSONNEL FOR SCANNING INSPECTION TASKS

Many inspection jobs require a person simply to look at quality characteristics to determine whether or not they meet quality standards. Jobs of this general type are usually referred to as scanning inspection jobs. Although certain minimum levels of visual acuity and mental alertness are probably required for inspection jobs of this type, little success has resulted from attempting to predict inspection performance from measures of visual acuity or mental alertness. There appears to be a relatively specialized aptitude or combination of aptitudes required for scanning inspection work.

A short test of inspection aptitude designed specifically for use in the selection and assignment of inspection personnel has been developed and validated [2]. The assumption that a paper and pencil task could be developed to incorporate the elements of typical tasks involved in scanning inspections underlined the development of the test. It was felt that this approach was the most likely to result in a reliable and valid selection method. The concept of the resulting test is described by the test instructions shown in Figure 11.3, which is the instruction sheet from the Harris Inspection Test (a commercially available form of the test that was developed).

Test Validity

Six validity studies involving a total of 90 inspectors were conducted, each within a different inspection department of Autonetics. Each department was responsible for inspecting a different type of product. In each study the relationship between inspection test scores in inspection performance measures was determined.

Performance measures were obtained in each study from actual inspections performed under controlled conditions. Each inspector in the study individually inspected the same item or set of items; his performance was measured in terms of percentage of defects detected. A master list of defects for each item was typically determined from inspections of the item by the inspection supervisor, a senior inspector, and a quality engineer. Those defects for which there was total agreement among this group comprised the master defect list. The performance of an individual inspector was expressed as the percentage of those defects he detected on the master list. The reliability of the performance measures was found to be relatively high. In all cases the correlations between inspection performance measures on alternate items were found to be greater than 0.80.

The relatively high and significant correlations between test scores and performance measures in each inspection department indicated that the validity of the test was not only high but relatively general for various types of scanning inspection tasks. Validity coefficients obtained in the six studies from rank order correlations between test scores and performance measures are shown in Table 11.1.

HARRIS INSPECTION TEST

FORM A

Developed by:
Douglas H. Harris

This is a test of how well you can inspect objects for defects. The task you are about to perform will require only five minutes. Please read the following instructions.

The enclosed area on the back of this sheet contains a number of objects. There are four different kinds of objects—one of each kind is shown in its correct form below. A correct example of each one is also shown on the back of the sheet.

Your task will be to inspect the enclosed area and put an "X" through each object which is not like the correct one shown.

Some of the objects may not have the correct shape, some may not be of the correct size, some may not have the correct number of parts or the parts may be out of place, others may be slightly out of line. For example, the object crossed out on the left below has a dot out of place. The crossed-out object in the center is out of line and the one on the right has two lines which are not properly joined.

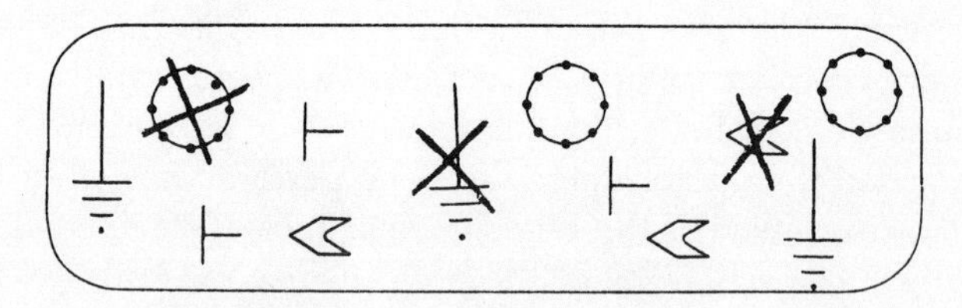

One object may overlap another. Overlapping is NOT to be counted as a defect. Your score will be based upon the number of defective items you mark. However, be careful not to mark any correct items—a penalty will be subtracted for each one marked.

You will have 5 minutes to complete your inspection.

Figure 11.3 Introduction sheet from Harris Inspection Test.

Table 11.1 Inspection Test Validities

Inspection Operation	Number of Inspectors	Validity Coefficients
Electronic chassis	11	0.39
Inertial instruments	8	0.86*
Module assemblies	19	0.58*
Circuit boards	27	0.64*
Microelectronic devices	10	0.78*
Photographic materials	15	0.52*

* Statistically significant beyond the 0.05 level.

To illustrate the general validity of the inspection test across the six different inspection jobs, the expectancy chart of Figure 11.4 was developed. This chart was prepared by identifying a group of 45 "superior" inspectors, those who were in the upper half of their department with respect to inspection performance. All 90 inspectors were divided into five equal groups on the basis of test scores. The expectancy chart was then formed by determining the percentage of superior inspectors within each range of test scores. The chart shows that the higher the test cutoff score, the greater the chances of selecting superior inspectors.

Test Reliability

Test reliability in this examination was found to be adequate. The reliability was determined by correlating scores on one form of the test with scores on the other form and using this correlation to estimate the reliability for the test. The reliability coefficient computed from test scores of 90 inspectors was 0.85.

The significant correlations between test scores and performance measures were not accounted for by the intervening effect of inspection experience, although a relatively wide range of experience—from a few months to a number of years—was represented by the participants in each study. Rank order correlations between test scores and months of experience were essentially zero in each study.

SELECTION OF PERSONNEL FOR MEASUREMENT INSPECTION TASKS

To investigate the selection of personnel for precision measurement tasks, a battery of personnel tests for the selection of machined parts inspectors was developed and evaluated [3]. To develop a battery of tests,

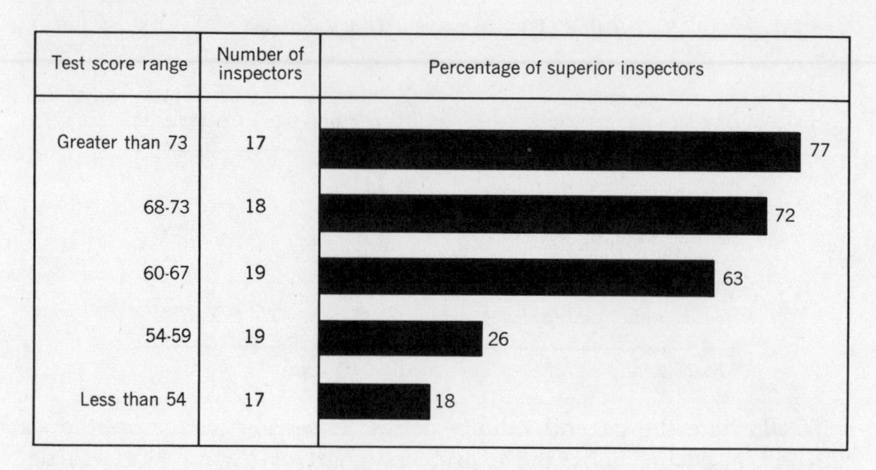

Figure 11.4 Expectancy chart showing (for 90 inspectors) the percentages of superior inspectors within each range of scores on the Harris Inspection Test.

the present job description for machine parts inspectors was reviewed, and the machine parts inspection job was analyzed to determine the skills and aptitudes required. Four major activities which could be sampled by short paper and pencil tests were identified. The results of this analysis are summarized in Table 11.2. No attempt was made to select tests covering every aspect of machine parts inspection; this would have been both

Table 11.2 Machined Parts Inspection Activities Sampled by Four Tests

Activity Performed	Test Selected
1. Compare obtained readings with drawings and specifications.	1. Minnesota Number Comparison Test Testing time: 8 minutes
2. Set up various precision measuring devices, requiring the use of basic mathematics, algebra, and geometry.	2. Purdue Industrial Mathematics Test Testing time: 25 minutes
3. Measure dimensions directly with various types of micrometers and gauges.	3. Can You Read a Micrometer? Testing time: 6 minutes
4. Determine the required dimensions of machined parts from engineering drawings.	4. Can You Read a Working Drawing? Testing time: 10 minutes

costly and impractical. The objective was to use small samples of critical inspection behaviors which could economically be obtained in the normal employment situation.

The rationale for matching each test listed in Table 11.2 with each machined parts inspection activity is presented below. Items from the four tests are shown in Figure 11.5.

1. *Number comparison.* All the measurements taken by the machined parts inspector must be compared with the corresponding dimensions in drawings or specifications. Therefore the number comparison portion of the Minnesota Clerical Test was selected for the trial battery.

2. *Industrial mathematics.* The machined parts inspector has to use simple mathematics, algebra, and geometry in setting up various types of precision measuring devices. The Purdue Industrial Mathematics Test was specifically designed to measure basic arithmetic and computational skills.

3. *Micrometer reading.* A relatively obvious skill requirement for machined parts inspection is the ability to use standard tools and gauges to make direct measurements of various characteristics. To sample this type of behavior the "Can You Read a Micrometer?" test was selected.

4. *Blueprint reading.* Skill in reading and interpreting drawings was the fourth area of activity sampled by the trial battery. The "Can You Read a Working Drawing?" test was selected to measure this ability. This test requires the subject to determine the dimensions of a machined part from a standard three view drawing.

Test Validities

Each of 26 experienced machined parts inspectors took the four tests and also performed machined parts inspections under controlled conditions. The controlled performance measures were based on approximately 12 hours of inspection time in inspecting a set of machined part job samples. The scores on the tests were then correlated with the job sample measures of inspection performance. Each test was found to be significantly correlated with the performance measures.

The four tests were combined into a test battery so as to maximize the correlation of the set of tests with the performance measures. The results of this analysis are shown in Table 11.3. As seen in the table, when the mathematics test was combined with the number comparison test, the validity increased from 0.62 to 0.76. These two tests had the highest correlations with the performance measures. A slight increase in the multiple correlation was obtained by adding the third test, "Can You Read a Micrometer?"; no increase was obtained, however, by including the fourth test, "Can You Read a Working Drawing?"

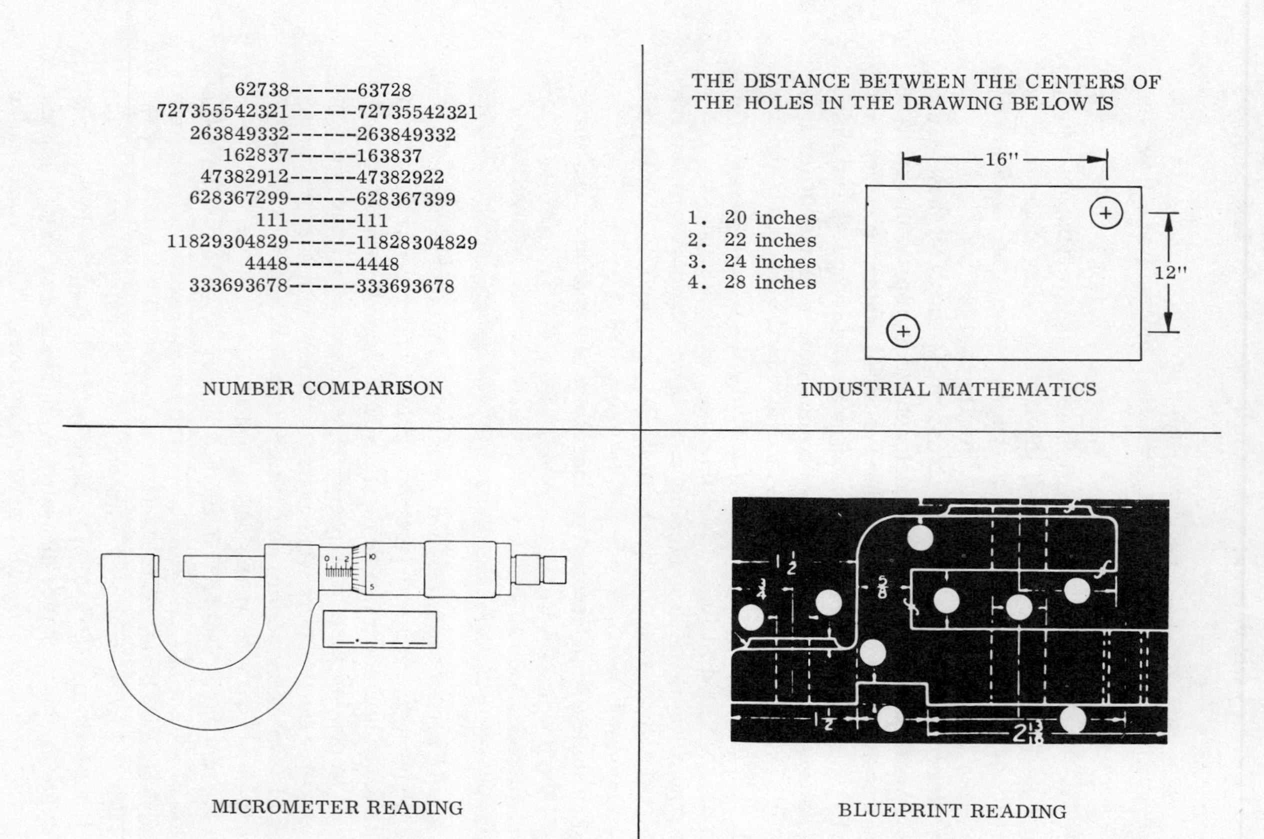

Figure 11.5 Sample items from each test of machined parts inspection activities.

Table 11.3 Multiple Correlations Between Measured Inspection Performance and Four Test Combinations

Test Combination	Multiple R
Number comparison	0.62
Number comparison + math	0.76
Number comparison + math + micrometer	0.77
Number comparison + math + micrometer + drawing	0.77

Test Value

To provide an illustration of the practical value of this level of prediction, an expectancy chart was constructed for a multiple correlation of 0.75. This chart was based on the method recommended by Tiffin and Vincent [4] and on the assumption that 50 percent of present employees were considered superior. As shown in Figure 11.6, the odds are 92/100 that an individual would be a superior inspector if his combined test score were in the upper 20 percent. In general, the chart shows that the chances of selecting a superior machined parts inspector may be greatly increased by using the combined scores for the number comparison and industrial mathematics tests. The more selective one can be in accepting applicants, the more valuable the tests would be.

Although the optimum level of prediction for this sample was obtained with only two tests, there are several reasons for using all four tests in the actual employment situation. The most important reason is that high scores on one test should not be allowed to compensate for lack of other

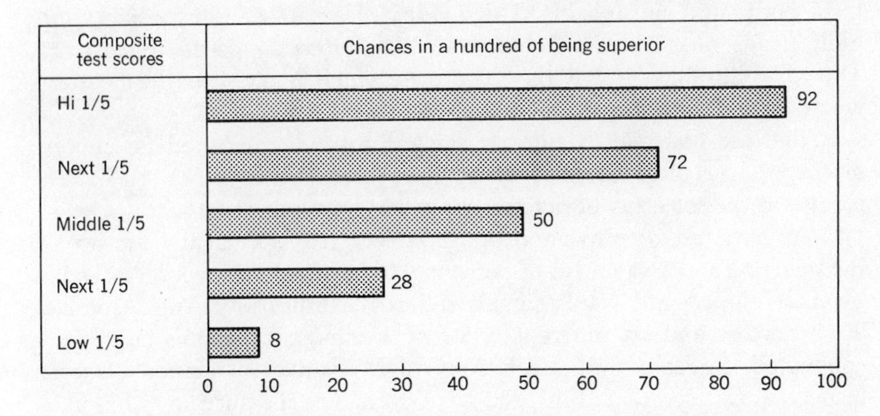

Figure 11.6 Expectancy chart showing the percentage of superior inspectors for various test score ranges.

critical abilities; certain minimum levels of all four abilities appear to be required for proficiency in machined parts inspection; for example, an applicant might be extremely high in number checking and mathematical ability but unable to read a drawing or a micrometer. To provide a practical method of using all four tests with few assumptions minimum acceptable scores were used with each test.

SUMMARY

1. Large differences have been found among people in their ability to perform inspection tasks. As a consequence the selection of people for inspection tasks on the basis of aptitude is likely to have a positive impact on inspection performance.

2. It should be realized that personnel selection methods seldom have the capability of assuring that the correct selection decision will be made every time; however, there has been success in developing and using selection methods to increase the "batting average" for making more accurate personnel selections.

3. Probably the most effective method to be employed in the selection of inspectors is the personnel test. A personnel test is a method used for obtaining a sample of behavior under controlled conditions. In using the information provided by a test or a series of tests for making personnel decisions, a person's test score is typically compared with the scores of others who have taken the test. If the person's score is in the range of scores of those people who have been successful on the job, the person would be assigned to the job.

4. To be useful in the selection and placement of personnel a test has to be both valid and reliable. The validity of a personnel test refers to its built-in design to measure objectively what it is supposed to measure. The reliability of a test is the degree to which it consistently measures whatever it is designed to measure.

5. Because there are frequently several aptitudes required for success on the job, it is often necessary to combine several tests into a test battery for use in the selection of personnel for certain types of jobs.

6. A short test of inspection aptitude designed specifically for use in the selection and assignment of personnel for scanning inspection tasks has been developed and validated. Six different studies involving a total of 90 inspectors and six different types of scanning inspection tasks were conducted. In each study a relatively high positive correlation between inspection test scores and inspection performance measures was found.

7. For the selection of personnel for measurement type inspection tasks, a battery of personnel tests was developed and evaluated. The test battery

developed to measure four different critical inspection behaviors included tests of number comparison, industrial mathematics, micrometer reading, and blueprint reading. The validity of the test battery was determined by relating test scores to controlled performance measures based on approximately 12 hours of inspection time of each of 26 experienced machined parts inspectors. The test battery was found to have a relatively high correlation with machined parts inspection performance.

REFERENCES

[1] Fred B. Chaney and Douglas H. Harris, Validation of Personnel Tests for the Selection of Machined Parts Inspectors. Paper presented at the Western Psychological Association Convention, Long Beach, April, 1966.

[2] Douglas H. Harris, Development and Validation of an Aptitude Test for Inspectors of Electronic Equipment. *J. Indus. Psychol.*, 1964, **2**, 29–35.

[3] Fred B. Chaney and Douglas H. Harris, Validation of Personnel Tests for the Selection of Machined Parts Inspectors, *loc. cit.*

[4] Joseph Tiffin and Norman L. Vincent, Comparison of Empirical and Theoretical Expectancies. *Personnel Psychol.*, 1960, **13**, 59–64.

XII

THE DEVELOPMENT OF KNOWLEDGE AND SKILLS

Of all the resources required for meeting quality assurance objectives, probably none is more critical than the knowledge and skills of people. Consider a quality system without human skills and knowledge. Clearly the system would cease to function; for example, quality engineering functions such as planning and problem solving would cease to exist and only the most elementary inspections, if any, could be performed. This, however, is a very crude examination of the need for knowledge and skills. The point is better made by viewing the magnitude and variety of skills and knowledge required for the effective attainment of quality assurance objectives. Since quality assurance is concerned with the entire product cycle, the quality system requires knowledge of customers, engineering, production processes, and purchasing operations as well as those operations that are unique to quality assurance activities. Because of the need to interact with people of different backgrounds and points of view, quality assurance typically requires communications and human relations skills in addition to certain specific technical skills.

Since people are not born with the skills and knowledge needed for quality assurance activities, it is necessary to consider how they can be developed. Further, because cost and timeliness are factors in the development of skills and knowledge, our concern becomes one of how these resources can be efficiently developed.

It is generally expected that people perform better as they gain experience on the job, but this expectation is reasonable only when there is a chance to learn from experience; for example, studies of inspection performance have shown that experience (time on the job) alone does not necessarily produce better performance. Measures of inspection performance obtained under controlled conditions usually indicate no relationship between amount of inspection experience and inspection accuracy [1]; that is, an inspector with 48 months of experience was not necessarily any better at detecting defects than an inspector with only two months of experience. Since there was ample room for improvement in all cases, this is an indication that most inspectors had learned little from their job experience.

How can knowledge and skills be developed within employees? An answer frequently given to this question is that knowledge and skills can be developed through training. Training refers to the set of formal procedures that a company uses to improve job related capability. Unfortunately the term training evokes an image of formal courses and classroom activity, a view that is much too narrow. In fact, chances are that classroom type activities may be one of the least efficient approaches to learning job related skills and knowledge. In this chapter the concept of training has been replaced by the concept of programmed experience. The central idea is that learning can be a continuing process that includes efforts on and off the job that have an impact on making a person better at his job. It is the responsibility of the supervisor to program an individual's development so that learning is most efficient. If classwork is deemed to be the most efficient way to learn certain aspects of the job, then classwork should be employed. If on the other hand, learning is likely to be more efficient if knowledge of the results of job performance is provided on a regular and timely basis, then a performance feedback system should be established for the job. The concept of programmed experience enables the supervisor to avoid the stereotypes and traditions generated by the concept of industrial training.

In the development of the skills and knowledge of his people the supervisor does not do everything himself. It is his responsibility to consult with appropriate specialists and with his people in the assessment of development needs and in the consideration of appropriate methods to employ. This chapter includes some of the basic information needed by a supervisor to meet his personnel development responsibilities.

THE OBJECTIVE: PERFORMANCE IMPROVEMENT

The development of skills and knowledge is but an intermediate step toward the ultimate objective of improving performance on the job. There is a temptation to treat skills and knowledge as end products rather than as means to an end. When this happens, learning tends to become inefficient because skills are developed which are not necessarily required for job performance and useless knowledge is gained while useful knowledge is overlooked. Thus it may be helpful at the outset to examine the relationship between performance and learning and to review the potential contributions of efficient learning.

Behavior Change Infers Learning

At this time it is not possible to directly observe the process of learning. Learning is inferred from observations of behavior made before, during, and after specified experiences have occurred. This can be illustrated by observing a person learning to ride a surfboard. During his first day in the water the individual appears incapable of even the most fundamental activities, such as sitting on the board. After repeated trials and some coaching from an experienced surfer the person is able to falteringly paddle the board through the water. However, there are many random movements that do not appear to assist the surfer in either paddling or guiding the surfboard. Weeks later the developing surfer is seen paddling to catch a wave, standing upright on his board for several seconds, and then splashing off into the water. Finally, the surfer is able to get his board into a wave and ride it to the left or to the right with some degree of ease and grace. Eventually, perhaps, the surfer may join the small fraternity of surfing enthusiasts who are able to hang five (wrap the toes of one foot around the nose of the surfboard while angling along the face of the wave) and perform other stunts with apparent ease. From these observations it is inferred that the person has learned to surf.

Careful observations of the progress of the surfer would probably indicate that learning took three different forms. First, some knowledge was gained; for example, the surfer learned where to place his body on the surfboard during different wave conditions. Second, the surfer developed a set of skills; he reduced a large number of uneven actions to a small number of purposeful motions. Third, some changes in attitudes occurred. Early in the learning process, there was evidence of apprehension and feelings of inadequacy, but after most of the elements of surfing had been mastered, feelings of confidence and pleasure were more prevalent.

The development of personnel in industry is often concerned primarily with acquiring new knowledge. Development of this type is really only communication with an emphasis on learning what is communicated. In quality assurance activities this aspect of learning may be concerned with the use of new operating procedures, specifications, test requirements, or inspection instructions. Learning is demonstrated by the effective use of this information or by correctly answering questions about the content of these materials.

The development of skills is directed toward improving a person's capability in performing a task. Although it is necessary to provide new information about how to do something, the development of skills also requires the types and amounts of practice required to master a task or to improve task performance. The development of different skills requires the use of different techniques; for example, efficiently developing skills in inspecting printed circuit boards may require an emphasis on practice with rapid feedback of performance accuracy while the development of skills in quality engineering may focus on the acquisition of knowledge followed by some supervisor coaching during application of this knowledge to quality engineering tasks.

The modification of attitudes is frequently integral to the development of skills and the acquisition of knowledge. Attitude changes may be required before any progress can be made in the other two types of learning; for example, because there can be little success in developing skills in someone who does not have a desire to learn, the initial part of skills development may be directed toward creating an attitude favorable to learning the required skill. On the other hand, attitude change alone may be the primary factor that leads toward improved job performance. The entire subject of motivation in industry is closely related to employee attitudes. The consideration of motivation in attaining quality assurance objectives is discussed in the following chapter.

Potential Contributions of Efficient Learning

There are two primary contributions to be made by increasing the efficiency with which learning takes place in a quality system. These are increasing the capability of the system and reducing system costs. It should be recognized that learning is a process that may be taking place continuously in the quality system. This learning, however, may have negative, positive, or zero effects on human performance in the system. Inspectors, for example, may learn that management is concerned with certain quality characteristics and then concentrate on the inspection of these characteristics to the exclusion of other equally critical characteristics. This may result in an overall decrease in inspection accuracy. On the other hand, as the result of learning some new analytical techniques, quality engineers may increase the effectiveness with which they identify critical quality problems. By the effective programming of experience, knowledge and skills are learned that increase the capability of the system for the attainment of quality goals and objectives. More efficient learning will result in greater system capabilities being developed in shorter periods of time.

Efficient learning may contribute to cost reduction in several ways. Personnel costs may be reduced by decreasing the time required to bring new employees to acceptable levels of job proficiency and by improving the productivity of experienced personnel. The costs of materials and supplies may be reduced by decreasing errors in job performance. Obviously increasing capabilities and reducing costs are directly related contributions. An increase in capability is nearly always reflected by a decrease in the costs of meeting quality objectives.

THE FIRST STEP: IDENTIFICATION OF NEEDS

Programming for efficient learning requires relatively precise answers to questions of what, who, and how. What skills and knowledge are required? Who is to be involved? How can learning be made most efficient? These questions are part of the basic problem of allocating system resources to meet quality objectives. Consequently both the problem of allocation of resources and the specific techniques for identifying learning needs are to be considered.

The Allocation of Resources

The management of quality assurance activities involves the allocation of people, equipment, and information. An important aspect of this allocation of resources is the effort directed toward obtaining, developing, and utilizing human resources. First, needs for human resources must be identified and then steps taken to satisfy these needs. In identifying the needs for skills and knowledge, it becomes apparent that needs exist at at least two levels; there are system needs and individual needs. The determination that a new production effort will require 40 persons skilled in the inspection of complex electronic modules illustrates needs at the system level. The determination that these inspectors require skill in detecting defects in welded bonds reflects needs at the individual level. Our concern in this chapter is primarily with developmental needs at the individual level. Discussions of the techniques to be employed in need identification and satisfaction are directed toward the development of skills and knowledge in individuals.

Techniques for Need Identification

The efficiency with which skills and knowledge can be developed in quality assurance personnel is a function of the precision with which developmental needs can be identified. The desired approach is one of specifying, in a systematic fashion with a relatively high degree of detail, needs that can lead directly to the programming of some type of experience. Several techniques have been developed and employed for doing this. Their effectiveness, of course, is highly dependent upon the type of job involved and whether the primary concern is with job knowledge or job skills. The basic types of techniques are described briefly below.

Task Analysis. A task analysis is a systematic way to list all the things that a person does on a job. In identifying needs for skills and knowledge, a task analysis ensures that no aspect of job performance is overlooked. The amount of detail included in the analysis may vary from job to job but it is generally desirable to use two or three levels of description; for example, we may talk about general functions to be performed, specific tasks, and detailed subtasks. Examples of these three levels of detail for a typical quality engineering job are

(a) function: monitor quality level,
(b) task: review quality information transmittal cards,
(c) subtask: check defect description on card

The resulting task list serves as a framework for identifying specific developmental needs for any work group. This, however, is just the starting point and other procedures must be used to measure the difference between existing and desired performance levels of specific tasks and to relate these differences to needed job knowledge or skill improvement.

Performance Measures. Usually the best method for identifying training needs is the analysis of measures of actual job performance. One type of performance measure, the inspection job sample, has proven to be an effective method for identifying, in detail, developmental needs in inspection job knowledge and skills. Inspection job samples were described earlier together with a step-by-step procedure for their application.

The job sample approach also can be adapted to other quality assurance jobs such as supervision and engineering; for example, to measure performance in handling program directives, internal correspondence, corrective action requests, drawings, and specifications, a job sample can be structured so that people are required to analyze these materials and make appropriate recommendations or decisions. The adequacy of their skills for each task can then be measured in terms of the percentage of correct decisions made and the time required to complete the task.

Job Knowledge Tests. Needs in areas which are highly dependent on knowledge, as opposed to skill, can be measured by objective job knowledge tests; for example, the effectiveness of both inspectors and quality engineers is highly dependent on their knowledge of specifications and operating procedures. It cannot be assumed that simply having these documents available for the individuals who use them will result in adequate job knowledge. Nor can it be assumed that reading and initialing new procedures and specifications is an effective means of maintaining the required level of job knowledge. There are probably very few jobs in quality assurance operations that would not benefit from the periodic use of objective tests to measure the amount of knowledge which employees have about critical activities. On the other hand, care should be exercised in the application of job knowledge tests to measure only those types of information that are actually critical to successful job performance; for example, employing tests to measure the knowledge of a procedure that is never used would not only be a waste of time but would also discourage the use of job knowledge tests for more legitimate purposes.

Information Checklists. In addition to general job knowledge, a number of specific items of information are necessary to successful job performance. Since this information may be highly variable and quite specific, it is not practical to learn it all; however, people must know how to find and use this information. The checklist is a practical tool for identifying training needs in the ability of people to find the information they require for their job. The supervisor may develop a checklist directly from the task analysis of the job. The checklist may then be used by the supervisor to determine training needs through interviews with employees.

Questionnaires. The questionnaire method of determining training needs may be used to collect judgments about the skills and knowledge areas or it may be used to measure attitudes which are believed to influence job performance. In all three areas questionnaires have the advantage of collecting data directly from individuals who will be most involved in the developmental effort, namely, the people to be developed. Asking people directly about their needs for skills and knowledge is particularly important because the value of the developmental effort will depend on the attitudes of these people. The employee who participates in a plan that he helped formulate will generally make a greater effort to take advantage of the opportunity for improvement than people who are subjected to the developmental effort without first being involved in its planning and implementation. Since this method is easily influenced by biases of one type or another, it is best used in conjunction with one of the more objective techniques.

FACTORS IN EFFICIENT LEARNING

Faced with the need of developing specified capabilities within an individual or group of individuals, it would be helpful to have at hand a set of principles of learning that could be applied to make the developmental process most efficient. Efficient learning could then be assured by programming experience for the individual or group in accordance with the set of learning principles. Unfortunately the existing theories and principles of learning do not appear very impressive to a person charged with the development of knowledge and skills in other people. All he has available to him are some general guidelines which are consistent with common sense and which leave the questions of method and technique for him to answer; perhaps it is the complex nature of human learning that causes this to be so. There seems to be no lack of learning researchers or theorists; hundreds of reports and books are written each year on the subject of learning. For whatever the reason, the practical problem of developing skills and knowledge in a quality system can best be attacked at this time by considering factors that have enhanced learning in other practical situations rather than through the application of a set of learning principles stemming from a theoretical understanding of the learning process. Several of the more significant factors in efficient learning are now discussed.

The Desire to Learn

The motivation of the learner is an important factor in the amount of learning that takes place and the extent to which this learning results in improved performance on the job. To understand the motivational factors which influence developmental effectiveness it is necessary to focus on the needs and goals of the learner. From a motivational standpoint development efforts may be viewed as a means of helping individuals obtain skills and knowledge they feel will be useful to them. Unfortunately this type of activity is often viewed as something done to make people more effective rather than as a process of helping people to help themselves. The same motivational factors which must be considered in developing skills and knowledge are also important on the job. Because a detailed presentation of employee motivation is provided in the next chapter, only a brief discussion of the part played by the needs and goals of the learner is given here.

Before an individual can benefit from a learning experience, the readiness to learn must be established. Learning, generally, will not take place unless a person can see how the experience will satisfy some of his personal needs; for example, many people have taken the same route to work each day for a number of years but are unable to recall even a small percentage of the things they have seen each day. This results from the fact that they have never had a need to learn and remember the many things which they pass every day. The same condition can exist in any learning situation where the participants do not feel that the experience will be of personal value to them. It cannot be assumed that just because the learning experience provided is needed and important for the organization that people will feel that it is worthwhile for them. In fact, it would probably be more correct to assume that most people have a natural reluctance to learn new things, especially when this learning is aimed at changing patterns of job behavior with which they have become accustomed. For this reason it is necessary to relate the objectives of any effort in the development of skills and knowledge to the individual's needs and goals.

In addition to being meaningful, goals should be obtainable. If the goals of a developmental effort appear to be unobtainable, people may refuse to become actively involved in the development process. Even when people do make a conscientious effort to achieve unrealistic goals, they may experience failure and become frustrated. This may result in actual decreases in on-the-job effectiveness and negative attitudes toward the job. A useful procedure is to establish relatively modest goals during the early part of the development process—goals that are well within the capabilities of learners. The goals may then be gradually increased as the development effort progresses. Some provision should also be made for letting people set their own goals for final levels of achievement. However, the supervisor or person conducting training should play an active role in revising goals if they appear to be unrealistic. Since people tend to be optimistic about their capabilities, it is also desirable to base goals on actual learning data which have been obtained from similar groups.

Building on Present Knowledge

New material should be related to the person's present knowledge. The development effort must start with recognition of the knowledge and skills that the person has already developed. Information is easier to learn and recall if it is related to familiar facts or concepts; new tasks are easier to learn when critical elements are tied to present skills and habits. It is also useful to have information on the general background and experience of the learner so that meaningful examples can be employed. Much of this background information may be obtained during the identification of developmental needs.

Maximum Participation

In general, learning is most efficient when the level of participation of the learner is greatest. The amount of learning that takes place while a person is just listening or watching is usually quite small compared to the learning that results from his actually doing something. The success of certain types of programmed instruction technique described later in this chapter illustrates this principle. Active participation also tends to increase the acceptance of the objectives of the development effort and maximizes the impact of the developmental efforts on actual job performance. A high level of participation and involvement is especially critical when the effort is aimed at increasing skills or changing attitudes.

Knowledge of Results

The one factor, above all others, that is necessary for efficient learning is providing for knowledge of results. This factor is central to the learning process. The primary ingredient required by an individual in improving his performance is knowledge of how well he is performing and of those aspects of his performance that require improvement. Could a basketball player improve his ability to score baskets if he did not know whether his shots hit or missed? Could an electronics assembler learn to make fewer errors without knowing the types and amounts of errors currently being made? Could a supervisor become more effective without having some awareness of his impact on his organization?

Probably one of the most overlooked opportunities for improving performance in industry is that of providing knowledge of results on the job. Although it is probably one of the most effective methods of improving job performance, provision is seldom made for the establishment of individual or group objectives and the timely feedback of progress toward these objectives. A common objection to providing knowledge of results is that it is too costly to do so. However, in most cases where the costs of providing knowledge of results have been compared with the resulting savings from improved performance, the gains have far outweighed the costs. The chances are, then, that providing knowledge of results is a matter of spending money to make money.

There are several considerations in providing knowledge of results that make the difference between significant improvements in job performance and little or no improvement in job performance. These considerations fall roughly into two categories—relevancy and timeliness. When feeding back performance results, information must be relevant to the desired job objectives, otherwise improvements in performance will not occur. Although it appears to be quite obvious, this consideration can be easily overlooked when establishing the mechanics of a performance feedback system. If less relevant aspects of job performance are emphasized by the feedback system, these are improved at the probable cost of some deterioration in performance on the more relevant aspects.

For any given task there is probably an optimal time interval between the performance of the task and knowledge of results. Feedback should not be so immediate that knowledge of results cannot be properly assimilated and applied to subsequent task performance, nor can the interval be so long that the relationship between task performance and the results of task performance is no longer obvious or meaningful. The primary problem, however, in providing knowledge of results is one of making the time interval short enough; for example, there does not now exist a method for providing timely feedback of performance results to inspectors on the job. A defect undetected by the inspector, even though it may be detected at a later time, will probably not be reported to the inspector in time to be meaningful. Thus the opportunity for inspectors to learn on the job is seriously curtailed by the lack of timely knowledge of results.

Recall and Evaluation

Learning is not complete until the new behavior or information can be reproduced by the learner. When a person is listening or reading, he may easily assume that he is learning all of the new inputs he receives. In addition, people responsible for conducting training courses frequently assume that learning has been accomplished when they have told the group something or have given them appropriate material to read. Chances are, however, that very little learning has taken place without some provision for the learner to recall the information he has been presented or for some demonstration and evaluation of the skill he has been practicing. The process of recall and evaluation can be a very rewarding part of the learning situation if experience is programmed to provide testing and feedback after short periods of instruction.

With respect to most manual skills, it is possible to provide continuous evaluation or discrete feedback of performance results at frequent intervals. If the desired performance can be clearly specified, then self-evaluation becomes a natural part of practice. Attempts to transmit facts and information require specific test periods to measure the extent to which this material has been retained by the learner.

In programming the learning of large quantities of highly specific information, active recall of this information should be tested after each meaningful unit is presented. The presentation-recall process should be repeated until that unit of information is completely mastered. As the material becomes more general and interrelated, the periods of testing will necessarily be spaced further apart, but this does not reduce the need for periods of active recall. In teaching even the most complex concepts at least 25 percent of the training period should probably be spent on active recall and evaluation.

Other Factors

The six factors described above do not exhaust the list of considerations in efficient learning. Research has identified a number of other factors which may influence the learning process. These include the way practice is distributed during the learning of a skill, the effect of rewards and punishment, the relationship established between trainer and learner, and how the task or information to be learned should be divided for the learning process. In the opinion of the authors there is insufficient evidence at the present time to provide clear guidelines with respect to these other factors. It is felt that discussing the pros and cons of available research findings would not contribute substantially to the attainment of more efficient learning in quality systems.

DEVELOPING THE LEARNING PROGRAM

The first concern in developing knowledge and skills within the quality system is with the jobs and tasks which are performed. To the extent that it is possible, the factors discussed above relative to efficient learning should be incorporated into the various jobs performed by people in quality assurance operations. Thus the first step in developing a learning program is one of determining the extent and the ways that jobs can be changed to more fully integrate the learning process with actual job performance. It should be recognized that this effort may actually require changes in the design of the job and in the way the job is related to other activities. The fact that many people feel that this is not an appropriate concern for a training effort illustrates how narrow our thinking has become about how knowledge and skills are developed within an industrial system.

After the system and individual needs for knowledge and skills development are identified and steps taken to enhance the extent to which this development can take place on the job, the supervisor can look into supplementary developmental techniques that can be applied in satisfying the developmental needs that cannot be met integral to the job. The supervisor has the responsibility for selecting and applying the techniques that will most effectively meet these needs. When extensive development is needed by relatively large numbers of employees, the most effective method may be formal, off-the-job courses where lectures, demonstrations, films, and other training aids may be employed. Formal classroom training is relatively well understood because most of us have spent the greater part of our earlier years undergoing training of this kind; therefore it is not discussed further here. The supervisor may also meet many developmental needs through the use of supplementary techniques which can be applied on the job. Since the potential of many of these techniques is often not fully realized, several approaches that have been demonstrated to be particularly useful will be described. These supplementary development techniques are supervisor instruction, coaching, self-instruction, and job sample instruction.

Supervisor Instruction

Probably the most common approach to supplementary on-the-job development involves instruction by the supervisor or his representative. When conducted in a systematic manner, supervisor instruction can have a significant impact on performance. This was illustrated in a study of machined parts inspectors [2]. Job sample performance measures were used to identify needs for the development of knowledge and skills for a machined parts inspection department. Three tasks were identified as primary inspection problem areas: detection of mislocated holes, thread gauging, and measurements of parallelism and concentricity.

Four one-hour sessions were conducted by the supervisor on basic measurement techniques in an effort to overcome these deficiencies. The material covered in each of the one-hour sessions is described briefly.

1. *Precision measurement.* This section was primarily concerned with the use of gauge blocks and/or wear blocks as calibration standards, fixed production gauges and height or thickness gauges. Proper set-up procedures and possible sources of error were discussed for each application.
2. *Elements of thread gauging.* The proper use of thread plug gauges, thread ring gauges, and thread snap gauges was discussed. Screw thread nomenclature and interpretation of thread symbols were also covered in detail.
3. *Internal comparator.* A detailed procedure for use of the Sheffield Model N-8 gauge was provided; the advantages and disadvantages of this technique were discussed.
4. *Interpretation of general drawing notes.* This session was primarily a question and answer period; examples of drawing notes which had been misinterpreted in the past were presented and discussed. The participants were encouraged to provide samples of problem areas from their current work assignments.

The effectiveness of this approach was evaluated with two groups of experienced machined parts inspectors. The groups were matched on measures of four aspects of performance which had been collected six months earlier. The members of the first group served as a control and did not participate in the training. The second group received the training outlined above. Performance for each group was measured on matched sets of machined parts with approximately 60 defects per set. The average inspection time for each set was 12 hours. Findings indicated that the supervisor conducted instruction resulted in a 32 percent increase in the number of defects detected. No significant change was obtained in the control group.

Coaching

A person proficient in the skills or knowledge needed by a work group may be employed as a coach under certain circumstances. This approach is most useful when the needs of individuals in the group are highly varied and it is preferable that the people not be taken away from their work for even relatively short periods of time; for example, an inspection supervisor may call upon an expert on soldering techniques if this appears to be an inspection problem. The expert, or coach, may then spend several days with the work group observing each inspector's work and conducting interviews to find specific problem areas. As he works with each inspector, he may provide specific information and demonstrate skills needed to improve performance. The approach may be somewhat costly since only one operator can be involved at a time; therefore it is not particularly efficient when large numbers of employees have about the same developmental needs.

Self-Instruction

The technique of self-instruction may be a particularly effective method of providing new knowledge to quality assurance personnel. The large volume of new procedures, specifications, and other types of job instruction material create a constant need for the acquisition of new knowledge. This kind of information is traditionally distributed in written form with the assumption that it will be retained by people simply through the process of their reading the material. Research has shown, however, that this is not a particularly effective method of acquiring knowledge and that little confidence can be placed in the comprehension and retention of information which is transmitted in this manner. Although several presentation methods and formats are possible, self-instruction is most efficient when materials are programmed in accordance with the following basic approach. This approach has become generally known as programmed instruction.

1. Information is presented to the trainee in small segments or frames.
2. Provisions are made for an active response to each frame and a measure of the individual's comprehension of each frame.
3. Immediate feedback on the correctness of each response is provided.
4. The individual sets his own pace for learning the material.
5. The material in each frame is of relatively low difficulty so that individual error rates are quite low.

Perhaps the most important features of this method are the active participation by the trainee and the immediate knowledge of results he receives at each step in the learning process. The method is also highly suited to the needs of a busy organization in that the training can be accomplished on an individual basis as time permits. There is no need to take people away from their job for extended periods of time.

One of the disadvantages of this technique is that good programming is a time-consuming process and requires the services of a skilled programmer of learning materials. As a result programming should probably be limited to those types of procedures and specifications which must be learned by a large number of individuals. Illustrative frames from a programmed instruction sequence on defect classification are shown in Figure 12.1.

Job Sample Instruction

The use of the job sample as an instructional technique is particularly effective with inspection personnel. The development and use of job samples as measures of inspection performance was described in an earlier chapter. An inspection job sample for use in developing knowledge and skills is prepared in essentially the same way. Use of the job sample permits the application of several of the factors in efficient learning. It actively involves the inspector in the learning process and makes it possible for each inspector to obtain meaningful and timely knowledge of results. In addition, it provides a continuous record of learning and makes it possible to objectively evaluate the effectiveness of the instruction procedure.

Use of job sample instruction was demonstrated in the development of photomask inspectors. The visual inspection of photographic masks used in producing circuit boards is an extremely difficult task because of the small tolerances involved. It is a critical task because undetected defects at this stage may result in costly scrappage of hardware at later stages in the production process. Therefore a study was conducted at the Autonetics Division of North American Rockwell Corporation to evaluate the effectiveness of job sample instruction as a means of increasing photomask inspection effectivity. The study also provided information on the relative importance of experience versus aptitude in photomask inspection performance. The training was introduced by giving the participants a list of defect categories and standards of acceptability for each category. After being shown samples of each defect type and given 30 minutes practice on a sample photomask they inspected a series of three photomasks with known defects and received knowledge of results after each mask was completed.

For purposes of evaluating the effectiveness of job sample instruction the 15 individuals who participated in the study were divided into three

(Information)

11-1

Now we know what a defect is. A good question to answer now is, "What will *happen* if something is different from a drawing or specification." Actually, we answer that question when we *classify* a defect. We say, "here's what we think could happen in the *future*."

(Response)

11-2

When we__________a defect, we are saying what we think could happen in the__________.

(Feedback)

11-3

classify
future

(Information)

12-1

Also, when we think about the future, we think of the *worst* possible outcome. We assume that the defect escapes to a *higher level* of assembly or to the field. So, we are making a prediction about the *worst* possible future outcome.

(Response)

12-2

We assume the__________possible future outcome when we classify a defect.

(Feedback)

12-3

worst

Figure 12.1 Illustrative frames from a programmed instruction sequence on defect classification procedures.

groups as follows: five experienced, five inexperienced with low inspection aptitude, and five inexperienced with high inspection aptitude. The high and low aptitude groups were defined in terms of scores on the Harris Inspection Test.

The results of the study indicated that job sample instruction produced significant improvement in the inspection effectiveness of the inexperienced personnel. The inexperienced participants with a high level of inspection aptitude performed as well after only three trials as did the experienced photomask personnel. The job sample instruction did not increase the performance level of the experienced personnel beyond that which existed prior to the instruction. These findings are illustrated in Figure 12.2.

As a result of these findings, it was concluded that photomask inspection proficiency of personnel with above average inspection aptitude can be brought to existing levels within three hours using job sample instruction. These conclusions were particularly significant in view of the feeling among inspection supervisors that it took many months to develop skills of this type and in view of the shortage of experienced personnel for anticipated expansion of production activities in this area.

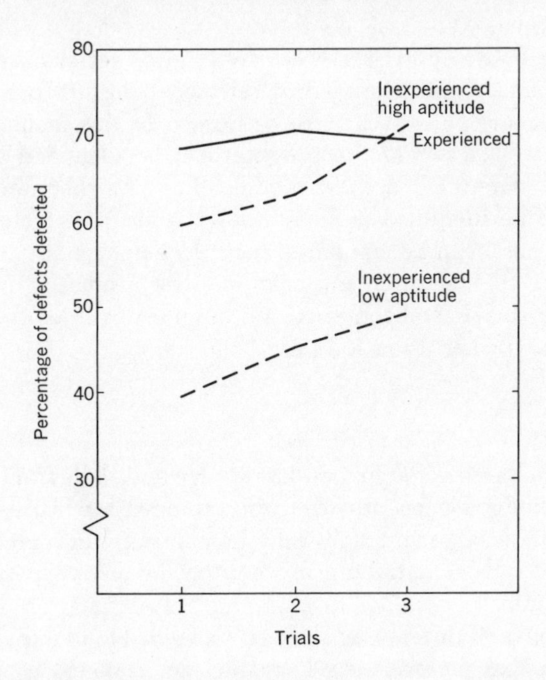

Figure 12.2 Effects of job sample training for three groups of photomask inspectors.

EVALUATING THE LEARNING PROGRAM

Just as the learner needs feedback with respect to his progress, supervision requires feedback with respect to the progress of the learning program. There are three primary reasons why the learning program should be evaluated. First, in a profit-making organization it is desirable to determine whether the gains of an effort outweigh the costs. Second, even if the program is more than paying for itself, chances are that further modifications and refinements can lead to even more efficient learning. Third, because a quality system is always changing, techniques which were once effective may become obsolete. Therefore, as the types of knowledge and skills required change, it may be necessary to change the approaches to personnel development. Thus the evaluation of the learning program is a matter of closing the loop. It takes us back to the beginning where it is necessary to consider new developmental needs and new approaches to the satisfaction of these needs.

The business of evaluating the developmental effort is primarily one of synthesizing information obtained about the progress of individual learners. Those measures which are made for the purpose of evaluating individual performance are basically the same measures that will be needed to evaluate the entire effort. Performance data for various aspects of the learning program can be averaged at selected points in time and compared with learning objectives at these points. In this manner potential strengths and weaknesses of the program can be identified and further investigated to determine desired modifications. Further developmental needs can also be identified and, by applying financial factors to these data, program costs can be compared with developmental gains in terms of dollars. Since the measurement techniques for evaluating performance in the quality system have been discussed in some detail in other chapters, they will not be further discussed here.

SUMMARY

1. People can be expected to gain knowledge and skills from experience only when that experience provides opportunities for learning. In considering how knowledge and skills may be efficiently developed learning should be viewed as a continuing process that involves experiences both on and off the job.

2. The objective of developing skills and knowledge in a quality system is improving system performance. Personnel development can contribute to system performance by increasing the capability of people in the system and by reducing costs associated with the system.

3. The first step in developing the human resources of a quality system is the identification of developmental needs. Needs for knowledge and skills exist at two levels—the system level and the individual level. System level needs refer primarily to the numbers of people required with specified capabilities. Individual level needs refer to the specific knowledge and skills required for an individual to be proficient on the job.

4. Techniques available for the identification of developmental needs at the individual level include task analysis, performance measurement, job knowledge tests, information checklists, and questionnaires. The effectiveness of each of these techniques is dependent upon the type of job involved and whether the primary concern is with job knowledge or job skills.

5. Before an individual can benefit from a learning experience, the readiness to learn must be established. In setting the stage for learning it is necessary to relate the objectives of the developmental effort to the individual's needs and goals. A person must be able to see how the experience will satisfy some of his personal needs. In addition, the objectives of the developmental effort should be challenging but obtainable.

6. Material to be learned should be related to the person's present knowledge. In doing this information obtained during the identification of developmental needs should be used in establishing the level of knowledge and skills that the person has already developed.

7. In general, learning is most efficient when the level of participation of the learner is greatest. A high level of participation and involvement is especially critical when the effort is aimed at increasing skills or changing attitudes.

8. Probably the most important factor to incorporate into programmed experience is knowledge of results. The primary ingredient required by an individual in improving his performance is knowledge of how well he is performing and which aspects of his performance require improvement. Providing knowledge of results is an effective means of increasing the opportunity to learn directly from job experience.

9. Since learning is not complete until the new behavior or information can be reproduced by the learner, opportunities should be provided for the learner to recall information he has been presented or to demonstrate the skill he has been practicing.

10. The first concern in developing knowledge and skills within the quality system is the possibility of redesigning jobs and tasks for the purpose of incorporating learning factors into them. Only after this approach has been explored fully should formal classroom training or supplementary developmental techniques be considered.

11. Supervisor instruction, coaching, self-instruction, and job sample

instruction are techniques that can be used to supplement learning directly from job experience. Each of these approaches has unique advantages depending upon the conditions under which learning is to take place.

12. To get the most out of a personnel development effort, provision should be made for evaluation. Measured changes in job knowledge, skills, or performance provide a basis for modifying and updating the development approach in light of identified deficiencies and new requirements.

REFERENCES

[1] James L. Thresh and John S. Frerichs, Results Through Management Application of Human Factors. Paper presented at the American Society of Quality Control Technical Conference, New York, June 1966.

[2] Frederick B. Chaney and Kenneth S. Teel, Improving Inspection Performance Through Training and Visual Aids. *J. appl. Psychol.*, 1967, **51**, 311–315.

XIII
QUALITY MOTIVATION

Motivation may be defined as the internal process that causes people to work toward goals that they feel will satisfy their needs. These needs may involve either basic biological processes or they may be highly dependent on learning and social values. Some of the most basic needs include hunger, thirst, and sex; on the other hand, needs for recognition, responsibility, and achievement are strongly influenced by learning and the values of society. The key to understanding motivation is the idea that people's needs direct their efforts towards certain goals or situations that they feel will satisfy these needs. As a result, motivation is clearly a matter of being attracted, not pushed, toward a goal or a set of goals.

Employee motivation is a very complicated subject as a result of the variability in people's needs and the variety of strategies that may be adopted to satisfy even the simplest needs. The study of motivation is also complicated by the fact that there are many other factors in addition to motivation that influence the behavior of people at work; the importance of these factors differs from person to person and from time to time. In spite of these problems, psychologists have developed some basic concepts that are useful in understanding employee motivation. The purpose of this chapter is to summarize these basic concepts as simply as possible and to show how they may be used by members of the quality organization to improve employee performance. To provide some concrete steps that can be taken to apply these concepts, a practical quality motivation program is described and its application illustrated.

NEEDS AND GOALS OF PEOPLE

Behavior occurs when people try to satisfy one or more needs. The relationships among needs, goals, and behavior are shown in Figure 13.1. The strongest need at any given time tends to direct the person's activity toward the goal or situation that he feels is most likely to satisfy that need.

If the goal is reached, then the resulting satisfaction should lead to a temporary reduction in the strength of this need. The relationships among needs, goals, and actions are usually most obvious when physiological needs are involved. The man deprived of food for an extended period of time will devote most of his effort toward obtaining food to satisfy his hunger. It is not difficult to understand why the physiological need for food leads to food-seeking behavior. Psychological and social needs, on the other hand, are usually much more subtle and difficult to understand. Often the person himself does not know the extent to which various combinations of psychological needs are influencing his actions, but because physiological needs are generally well satisfied psychological and

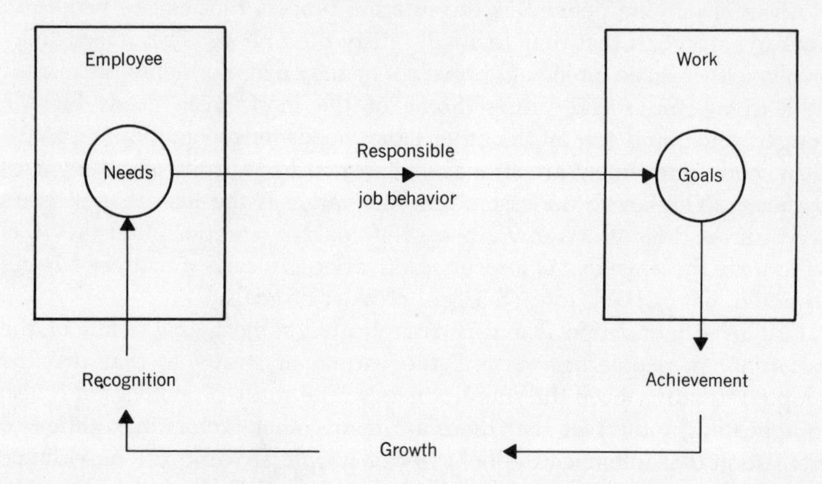

Figure 13.1 Relationships among employee needs, job behavior, and goal achievement.

social needs are the ones that are of greatest importance in industrial performance.

Herzberg [1] has developed a theory which provides a useful framework for describing the needs of industrial employees. This theory, which has been termed the "motivation/maintenance" theory, provides a two-way classification of the needs that influence employee motivation and job satisfaction. As shown in Figure 13.2, the first category contains maintenance needs and those aspects of the work environment that are not directly related to the job itself. Maintenance needs involve orientation, security, status, social factors, physical surroundings, and economic benefits that are not directly related to merit or performance. The theory suggests that these factors can serve as sources of dissatisfaction if they

Table 13.1 Principles of Job Enrichment

Principle	Motivators Involved
a. Removing some controls while retaining accountability.	Responsibility and personal achievement
b. Increasing the accountability of individuals for own work.	Responsibility and recognition
c. Giving a person a complete natural unit of work (module, division, area, and so on).	Responsibility, achievement, and recognition
d. Granting additional authority to an employee in his activity; job freedom.	Responsibility, achievement, and recognition
e. Making periodic reports directly available to the worker himself rather than to the supervisor.	Internal recognition
f. Introducing new and more difficult tasks not previously handled.	Growth and learning
g. Assigning individuals specific or specialized tasks, enabling them to become experts.	Responsibility, growth, and advancement

tion. It is also recognized that effective supervision requires a flexible type of leadership. The effective supervisor should be able to identify the critical factors in a job situation and adjust his leadership style to meet the needs of his particular group. Group participative techniques such as group problem solving and group goal setting were recommended as a means of obtaining deeper employee commitments to company goals.

One of the most critical aspects of a supervisor's job is the process by which he assigns work and evaluates results. Traditional forms of supervision emphasize the importance of clearly defined assignments and rely on some type of performance appraisal system for periodic evaluation. Performance appraisal systems generally involve some combination of structured interview and formal ratings. Under these conditions the supervisor views his job as one of identifying the employee's strong points and indicating areas where improvement is needed. A number of studies have shown that this type of appraisal system is not an effective means of improving performance [7,8]. The positive effect of praise presented in the appraisal situation is usually counteracted by the suggestions for improvement that are taken as a form of criticism. This criticism tends to increase the employee's defensiveness and leads to a breakdown in communication between the subordinate and his supervisor.

Huse [9] has shown that an alternative procedure, which does not involve formal ratings, is a much more effective way of improving performance. The procedure is a variation of management by objectives and has been termed work planning and review. The primary objective is to help the employee improve his work performance by establishing a more effective relationship with his supervisor. Work planning and review

are not maintained above certain minimum levels, but that they cannot be used to create high levels of motivation.

The second category includes motivation needs such as achievement, earned recognition, responsibility, and opportunity for personal growth. Herzberg feels that these needs which are directly related to the job itself are the most potent source of motivation. In other words, the theory suggests that the most effective way of motivating employees is to give them jobs that challenge their capabilities and to provide rewards that are tied directly to their accomplishments. The process of changing the

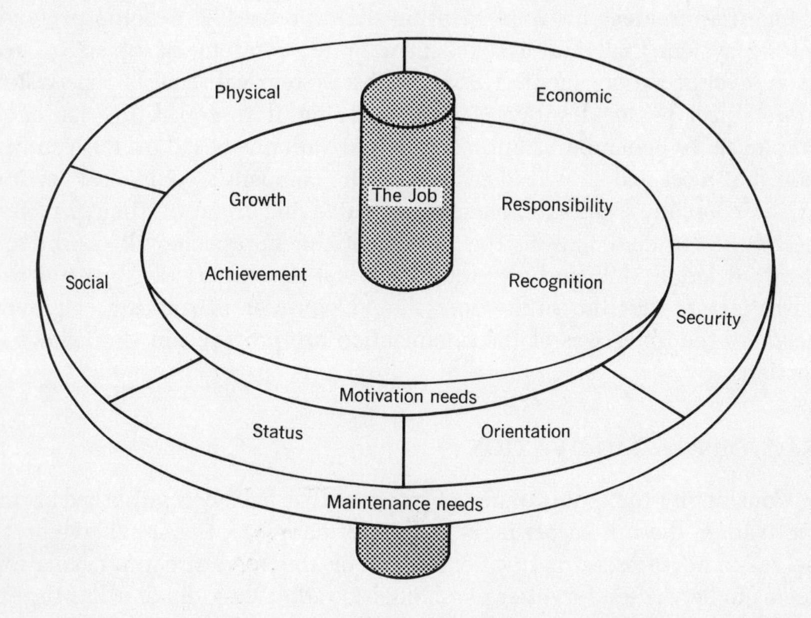

Figure 13.2 A classification system for employee needs (adapted from M. Scott Myers, Texas Instruments, *Harvard Business Review* (January-February 1965).

job to provide greater opportunities for responsibility, achievement, recognition, and personal growth is called job enrichment [2].

While the theory provides a useful framework for classifying these factors that influence job satisfaction, it appears to oversimplify the complex relationship between employee needs and motivation. A review of recent research in this area indicates that both motivation and maintenance needs can influence job satisfaction [3]. When a job gives employees a feeling of growth and accomplishment, they may remain satisfied even though maintenance factors such as fringe benefits, company policies, and even wages are felt to be inadequate. On the other hand, employees may become very dissatisfied even though they have satisfactory pay, supervision, and working conditions, if the job does not provide a feeling of meaningful achievement and recognition.

The concept of goal setting provides a framework for relating both motivation and maintenance needs to the objectives of the organization [4]. Positive job motivation exists when an individual is working towards goals that he thinks he can obtain and feels will satisfy his needs. In this framework the process of motivating people becomes one of helping them to establish goals that are related to their own needs for growth, recognition, achievement, and responsibility.

One requirement for implementing this approach is a company goal-setting system that establishes a meaningful hierarchy of objectives for each level of management. This type of a system will enable supervision to establish personal subgoals with meaningful responsibility for each employee. In general, this approach to motivation is based on the assumption that most people will identify with the company's goals, or a portion of the company's goals, as a means of satisfying some of their personal needs. In the final analysis the success of this approach will depend on the knowledge, skill, and attitude of the first line supervisor. Some of the basic factors that the supervisors should consider in relating employee needs to the objectives of the organization are provided in the following section.

FACTORS IN MOTIVATION

Considering the large volume of material that has been published about motivation, there is surprisingly little information on the specific steps to be taken to *improve* motivation. Most of the literature on motivation seems to be written for other psychologists rather than the first line supervisors who bear the primary responsibility for implementing motivation. Even if the supervisor has just returned from a one-week course in employee relations, he will usually have trouble in defining the specific motivational steps he should take to improve the quality and output of his group's effort. Does he have to change the job itself to motivate his people or is he the problem? Should he be more friendly and spend more time talking to his people? Should he leave more of the decisions up to his people rather than making them himself to save time? Is it a good idea to tell people how they are doing all the time or should he try to overlook the failings of some of his faithful employees? After reflecting along these lines for awhile, the supervisor may decide that this human relations stuff is not really that important. "After all, if an employee has any loyalty to the company he will always do his best, regardless of the

type of supervision, on the job he is given." Finally, the s
decide that the real problem is that the people just do n
recognition and reward for their accomplishments. "You
people to do a better job if you're not willing to pay for it.'

Fortunately research is beginning to provide objective an
of these questions. The conclusions from a number of maj
summarized below. It is hoped that this information will
practical guidelines for the busy supervisor.

The Job

One of the most obvious approaches to improving job sa
employee motivation is to change the job itself. For more t
efficiency experts of one type or another have been breakin
simple tasks to increase efficiency [5]. Throughout this p
ciency experts have met increasing resistance from the so
who have argued that the economy gained through special
because of boredom, job dissatisfaction, and decreased mo
argument has resulted in a counter trend towards job en
general, research has shown that job enlargement leads to
of motivation and productivity for supervisors, white collar
most blue collar workers in rural locations, but job enlarge
been particularly effective with blue collar workers in ur
Herzberg [6] has emphasized the importance of "job enri
provides an opportunity for increased responsibility and
growth as opposed to job enlargement that merely increase
of tasks performed. Recent studies in industry have resulted
ciples of job enrichment. These principles and the motiva
are summarized in Table 13.1. While a job which is ch
satisfying goes a long way toward assuring high levels o
there are a number of situations where job enrichment is
Under these conditions a supervisor must examine other fac
be used to improve the performance of his people. Even in
which job changes are instituted the supervisor will also wa
the impact of his behavior and attitudes on employee motiv

The Supervisor

A high level of employee motivation is one of the re
supervisor receives for doing his job effectively. In Ch
supervisor's leadership style was stressed as an important
ployee motivation. In general, the most effective supervisor
to be above average in both initiating structure and employ

involves periodic meetings between the worker and his supervisor for mutual determination of job goals, detailed planning of work to be done, mutual review of progress, and joint cooperation in solving job problems. There are no fixed times for the discussions but they usually take place at the beginning and end of projects. The discussions are confidential and no copies or records go to the personnel department. Hughes has provided a detailed description of the techniques used to successfully implement this approach in a large international corporation [10]. One of the critical steps in this implementation process was the use of individual coaching and counseling with each supervisor. At Texas Instruments the classic rating scale has also been replaced by a goal-oriented performance review. The system provides for joint goal setting, a review of accomplishments, and determination of pay status every six months. Numerous examples could be cited in which similar approaches to job planning and performance evaluation were found to be successful. One of the keys to the success of all management by objectives systems is the level of employee participation that is achieved during the planning process.

Employee Participation

In our earlier discussions of the supervisor's role in quality assurance examples were provided to show how motivation may be increased by helping employees to participate in the decisions that affect their work. The importance of using participation to relate job goals to employees' needs for achievement and recognition was also stressed. One of the most effective ways of accomplishing this objective is to allow employees to play an active role in decisions that affect their job. There are four levels at which employees may participate in decision making. From the highest to the lowest level, they are the following:

1. Actually making the decision.
2. Providing recommendations to those who will make the decision.
3. Learning from the decision maker the alternatives being considered.
4. Obtaining information about the decision after it has been made.

Motivation will generally be the greatest if the employees are allowed to participate at the highest possible level, but many of the decisions that are made about employees' work must necessarily be made by other individuals; these people by virtue of their positions of greater responsibility have more relevant information. Even in these situations, however, it is possible to create some feeling of participation by asking for recommendations and by clearly communicating the reasons for final decisions. This process is especially important where the effectiveness of the decision depends on its implementation by the employee group. One of the surest ways to kill motivation is for the supervisor to make arbitrary decisions and to communicate them to employees without any rationale or explanation!

It is important to realize that the use of participative methods does not imply any loss of management control. Unfortunately, participation is often associated with permissive or weak management. To the contrary, recent studies have shown that participative management creates stronger supervisors. They are stronger in the sense that they must be able to develop clearly defined goals for their operation and valid ways of measuring progress towards these goals. They are also stronger in that they have a more detailed understanding of their operation and its problems. Finally, they are stronger because they are more likely to provide their people with the type of support needed to meet their objectives. If participative methods are correctly applied, the limits of participation will

Participation increases motivation

be clearly defined and the situation will be structured so that the supervisor in no way relinquishes control of his employees.

The value of participation in reducing resistance to change has been demonstrated in an experimental study by Marrow [11]. Three levels of participation were compared during a change in job procedures. One group served as a control and received only an explanation of the reasons for change. Under the second condition, a moderate amount of participation was employed; the groups chose representatives who actually made decisions about the job changes. The third condition involved the high level of participation with everyone being directly involved in designing the new job. The high participation group was more productive during the change period and achieved a higher level of performance during the 30 days following the change than either of the other two groups. These findings are illustrated in Figure 13.3. When participation is used either as a means of increasing the employee's commitment to goals or as a means of increasing acceptance for new job procedures, the employee becomes more interested in information about how he is doing. As a result, feedback on job performance becomes an extremely important part of motivation.

Knowledge of Results

As was pointed out in Chapter XII, it is essential to provide knowledge of performance results in a regular and timely manner in order to increase and sustain high levels of motivation. Regular knowledge of results has proven to be of value from both informational and motivational points of

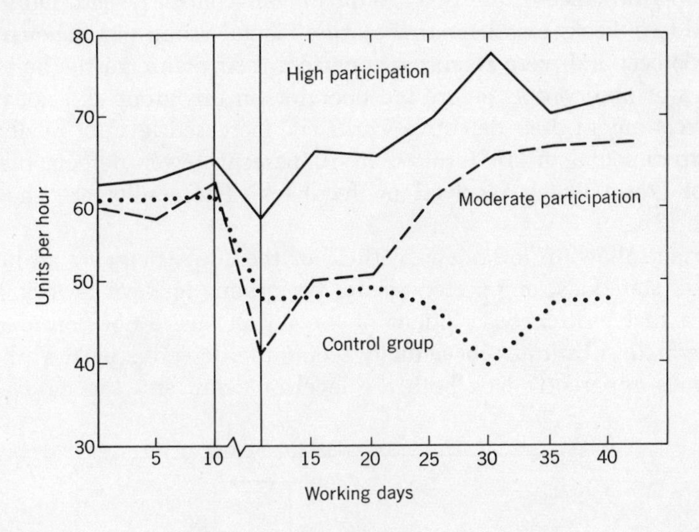

Figure 13.3 Relationship between participation and performance, adapted from (A. Marrow, *Industrial Psychology Pays in This Plan, loc. cit.).*

view. Before any improvement in performance can be expected, an employee must be informed of his past performance and be given specific suggestions for the kinds of improvements that are expected. In addition, feedback which indicates an individual is performing up to expectations, or is improving, can be an excellent means of sustaining motivation. Even when feedback indicates that performance is not up to expectations, it can provide an opportunity for employee motivation if the supervisor approaches the situation with a positive attitude. One means of accomplishing this is to indicate confidence in the employee's ability to achieve the desired goal and to concentrate on determining what type of assistance the employee needs to raise his performance to the desired level.

Although the value of feedback has been demonstrated in a number of laboratory and educational settings, only a limited number of studies has investigated the impact of knowledge of results in an industrial setting. Several recent studies at Autonetics, however, have confirmed the value of feeding back performance data to assembly personnel. One study investigated the extent to which feedback of defect information would improve performance of electronic assemblers [12]. A control group of seven operators and an experimental group of 13 operators were matched in terms of previous experience and prior performance level. Both groups were observed for a two-week period during which neither group was provided with a significant amount of feedback. There was no difference in the performance of the two groups for this control period. During the second two weeks members of the experimental group were shown all of their defects and were required to correct their own errors. The control group was observed as before but operators in this group did not review or correct any of their defective work. The increased level of feedback in the experimental group resulted in 70 percent fewer defects than the control group which received no feedback. The results are shown in Figure 13.4.

During follow-up interviews with 15 of the 20 participants all but one operator stated a strong preference for performing his own rework. It was learned that written descriptions of the defects were not considered to be adequate substitutes for actually seeing the defective part. Verbal descriptions apparently lack both the level of detail and the motivational

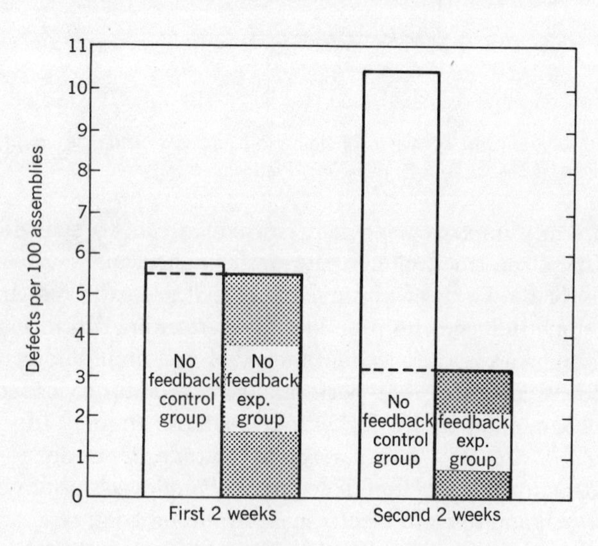

Figure 13.4 The effect of knowledge-of-results on electronic assembly performance.

value that is obtained from seeing and reworking the actual hardware defect. As a result of these findings and experience gained from conducting similar studies, the following recommendations were provided for effective feedback systems: (a) when a defective part is identified, it should be returned as soon as possible to the responsible manufacturing supervisor or leadman, (b) the supervisor should discuss each defect with the operator who produced it and make a specific recommendation for avoiding such errors in the future, (c) each operator should be allowed to perform his own rework, and (d) to provide a means of controlling feedback, random samples of rework items should be examined each week to determine what percentage is *actually* being returned to the original operator for rework.

In addition to rapid feedback of information on performance, some form of recognition for achievement is required to sustain high levels of motivation.

Recognition for Achievement

Perhaps one of the easiest but most neglected means of sustaining motivation is to provide recognition for achievement. Unfortunately many supervisors are so busy that their only contact with employees usually involves areas where improvement is required. Even when constructive criticism is provided, a continual pattern of critical comments rather than recognition for accomplishment will lead to decreased motivation. Recognition is a powerful form of motivation because it provides tangible evidence that extra effort is really worthwhile. Recognition is particularly effective when it is provided in a form of more interesting assignments, increased responsibility, deserved promotions, and merited pay increases.

One of the most essential steps in assuring effective recognition or reward practices is for the company to make sure that the rewards provided are really the ones that are most generally desired by employees [13]. Although this seems like an obvious and simple step, few companies have used techniques such as modern attitude measurement to improve the effectiveness of their reward practices. As a result many of our recognition practices may seem to be highly valued by management but have little if any effect on the average worker. The systematic determination of optimum reward practices is especially important in developing and administering employee pay practices.

Economic Incentives

The importance of money as a source of motivation depends largely on what the money represents to a person; money should not be regarded as a universal motivator. One of the primary problems in understanding the

value of money is that it is a symbol that represents different things to different people. It can serve to satisfy needs such as achievement, prestige, or security because it can be a measure of success, status, or protection. However, employees often leave higher paying jobs for others that are more interesting and provide a greater challenge. Other individuals may choose to remain in a secure low-paying job rather than risk an increase in pay with more responsibility and more chance of failure. As a result, we must conclude that the effect of money is highly dependent on what it represents to each individual.

In general, salary increases which are based on merit and other forms of pay which symbolize achievement have a high degree of motivational value. However, general pay increases, group insurance plans, and many other types of standard economic benefits which apply equally to all individuals in a company have little, if any, long-term motivational value. In fact, any type of economic benefit that becomes automatic or tends to increase at regular intervals is of little motivational value and may easily become a source of dissatisfaction if employee expectations are not met.

An effective technique for using money as a motivator is to tie salary increases to the accomplishment of some specific objectives. In this system the supervisor and the employee jointly develop the subordinate's objectives for the next six months or year. It is extremely important to both parties to agree on exactly how progress toward these goals is to be measured and what types or amounts of reward are to be given for various levels of achievement. At the end of the agreed-on period both men should jointly participate in determining the progress and establishing new goals for the next period. Unfortunately, this type of practice is generally limited to management and sales positions. Most first line supervisors are severely restricted in terms of the freedom that they have to determine type and amount of rewards that their employees will receive.

Because of the factors outlined, we should not rely on financial rewards as a primary means of motivating employees. Effective pay practices, however, can be used in combination with the other factors outlined above to produce substantial improvements in performance. A detailed approach for using this information for achieving and sustaining a high level of quality motivation is outlined in the following section.

AN APPROACH TO SUSTAINED QUALITY MOTIVATION

The preceding discussion of the needs and goals of people and the factors that influence motivation has set the stage for describing a specific approach that may be used by a typical supervisor to increase employee motivation. There is a continuing need for a means of turning the abstract ideas behind employee motivation into some concrete action that can be taken to improve the quality of employee performance. Recognition of this need by government and industry has created a great deal of interest in company-wide motivation programs as a necessary step toward improving product quality. Many of these programs have resulted in positive gains in performance, but most have been unable to sustain these gains in performance over extended periods of time [14]. This state of affairs appears to result frequently from a combination of the following factors:

1. Motivation program activities are often directed by a central staff function instead of being integrated into normal management procedures.
2. Attempts to improve motivation usually have not given adequate consideration to such factors as tools and job instructions that influence both motivation and performance.
3. Quantitative data have not been collected to provide a basis for evaluating alternative techniques and maintaining the required level of management support.

The approach used by Texas Instruments provides one example of a successful motivation/performance improvement program that has considered the above factors [15]. The combination of motivation seminars for all levels of management, work simplification efforts, goal-oriented performance reviews, and annual improvement surveys provides a comprehensive approach for implementing and monitoring a motivation program. It places the responsibility for employee motivation squarely on the shoulders of supervision and provides an environment for employee participation in job changes that make major contributions to sustained performance improvement.

Although the work at Texas Instruments and other organizations has identified the basic elements needed to improve employee motivation [16], the specific procedures to be followed in implementing an effective program are still in the developmental stages. The final goal of this developmental process is to achieve permanent changes in management, supervisory, and employee behavior that will sustain high levels of employee motivation/performance.

One of the most useful new developments in the study of management is the increased emphasis on collaboration between behavioral science and technology [17]. This approach requires a joint consideration of engineering and psychological factors that affect industrial processes. The concept is to establish *sociotechnical systems* that enable employees and technology to produce the best results. The approach described in the following sections was applied by the Human Factors Department at Autonetics over an 18-month period in an attempt to translate the concept of a sociotechnical system into practical management procedures. The work was conducted by individuals with training in both the physical sciences and psychology, in conjunction with experts in manufacturing and quality control management.

Since the results of a motivation program should be evaluated in terms of actual output, the approach was based on a consideration of the relationship between motivation and other factors that directly influence performance. Motivation was defined earlier as an internal process that causes people to work toward goals that they feel will satisfy their needs. *Motivation* is clearly a matter of *being attracted,* not pushed, *toward a goal.* As a result, attempts to improve employee motivation require the development of job situations that will enable employees to satisfy their needs and to participate in decisions that affect their work. In this framework the job environment and the supervisor's relationship with his people may be just as important as parts and tools in determining the output of a work group.

In addition, the job situation must also provide adequate methods, tools, and information before motivation can result in effective performance. In the chapter on supervision we found that the most practical techniques for improving performance is the use of work groups as a source of ideas for improving the job; for example, at Texas Instruments the use of employee teams to develop better methods resulted in savings of more than $1,000,000 during a one-year period [18]. Employee participation to obtain technical job improvements has three results that clearly illustrate the important relationship between motivation and performance.

1. Many direct and permanent improvements in performance can be obtained by using the work groups' ideas for better tools and methods.
2. A deeper commitment to achieving job goals is obtained when employees have an opportunity to participate in the solution of their job problems.
3. The extent to which a supervisor takes action and provides feedback

on the problems identified by his people is a critical factor in achieving high levels of motivation.

Based on these concepts of motivation and performance, the following approach was developed.

Performance Measurement and Feedback

The initial step in the approach is one of surveying the adequacy and availability of existing data on human performance. To be adequate performance information should be useful in identifying performance problems, evaluating proposed solutions to problems, and providing individual employees with knowledge of the results of their performance. When gaps exist in the data required for these purposes, changes have to be made in the data system. The most difficult performance data to obtain is that used to provide knowledge of results to individual employees. A workable feedback system for this purpose must meet the following three criteria: (a) detailed information must be provided to individual employees, (b) information must be provided in a timely fashion, and (c) information must be related to established goals which are meaningful to the work group.

Supervisory Development

The objective of this process is to obtain the changes in supervisory behavior and attitudes required to implement participative techniques in selected manufacturing work groups. Specific activities include management development seminars, supervisory orientation, employee group applications, individual coaching for each supervisor, and management evaluation sessions. This process is illustrated in Figure 13.5. To create some acceptance and support for the program two management seminars of one hour duration each are held with the supervisors and general supervisors of the participating departments. The basic concepts to be presented to the assistant supervisors are reviewed and suggestions are obtained for types of group participation that should be most advantageous for the specific departments.

Initial Training. The supervisors complete a series of four one-hour orientation sessions in which the basic principles of group leadership, problem solving, and goal setting are introduced. An informal group discussion method is used during these sessions to illustrate the effective use of participative techniques. The instructor uses discussion and questioning to help the trainees understand and accept the basic principles of participative management. On completion of the first four training sessions each assistant supervisor schedules a minimum of eight meetings with one of his employee work groups.

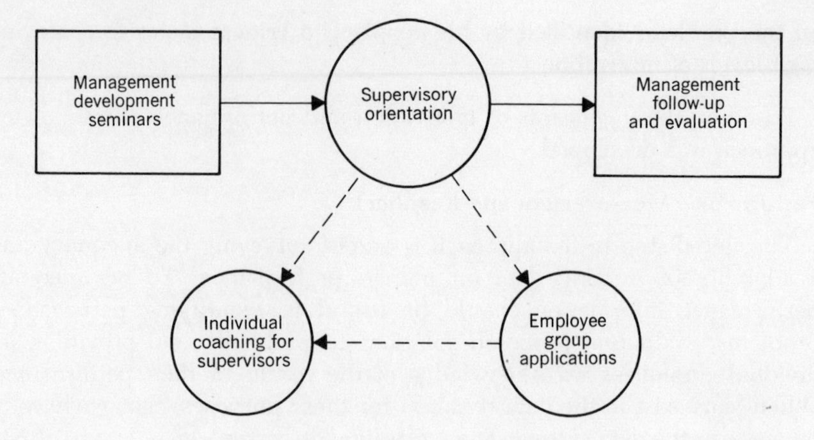

Figure 13.5 The supervisory development process.

Employee Group Applications. The general purpose of the series of supervisor-employee meetings is to increase the level of participation by the use of group goal-setting and problem-solving techniques. The specific objectives of the supervisor are as follows: (a) to obtain open two-way communication between himself and hourly personnel, (b) to obtain a high level of group identity or teamwork within the group, (c) to obtain a clear understanding of current performance levels and methods of measuring this performance, (d) to establish and obtain group commitments to future goals, (e) to involve the group in the identification and selection of the problems that must be solved to reach future performance goals, and (f) to obtain a better understanding of the relationship between the group's output and the final end product. A psychologist with experience in the group process attends each of the group meetings.

Individual Coaching. The supervisors receive individual coaching before and immediately after their group meetings. The psychologists who participate in the training feel that the majority of the changes in behavior and attitudes occur as a result of this individual coaching. Although it is extremely time consuming, it appears to be an essential part of the supervisory development process. The total time required for individual coaching varies from about 30 minutes to one hour per group meeting. Both the strengths and limitations of the supervisor are discussed. Rapid follow-up on problems identified by the group and the supervisor's role in implementing problem solutions are emphasized as critical factors in the success of the group process.

Goal Setting and Feedback. After a minimum of eight one-hour group sessions have been completed, the assistant supervisors are given two additional hours of instruction on the establishment of group goals and feedback. The objective of this training is to re-emphasize the importance of obtaining a high level of commitment to manufacturing standards for output and quality. As a result of this training the assistant supervisors hold several additional meetings with their employee groups to clarify and reemphasize the importance of their goals. In areas in which the goals are already established and achieved through the use of group techniques these meetings are used to express management's appreciation for the group's cooperation and to discuss the job factors that are critical in sustaining high levels of performance.

Follow-Up and Evaluation. The initial meetings are attended by a psychologist to assure that accurate records of job problems are maintained and to provide specific suggestions for increasing the effectiveness of the group meetings. The problem areas will also be reviewed by the psychologists to identify needs for research on new tools, visual aids, and methods. When significant improvements appear to be possible, the required research is conducted.

After the training and 10 to 12 employee group meetings are completed, monthly evaluation sessions are scheduled with the participating supervisors and appropriate levels of management to assure continuing application of the approach in areas where performance improvement is required. These meetings are generally continued for approximately six months or until the procedures become normal management practice.

QUALITY MOTIVATION RESULTS

The above approach has been implemented relatively widely in manufacturing and inspection operations at the Autonetics Division of North American Rockwell Corporation. As part of this implementation program, quality performance measures have been obtained and analyzed for selected groups, and controlled studies have been conducted to investigate aspects of the program. These findings have provided additional information concerning the relationship between work group participation and quality performance and have indicated factors to be considered in the application of group participation methods.

The initial studies were conducted in inspection operations to compare the effectiveness of group versus individual problem solving. Based on the inspection findings, a second series of studies were conducted in manufacturing to evaluate the combined effectiveness of group problem solving and goal setting. These studies were reported in the ASQC Quality Motivation Handbook [19]. The results and conclusions for the research in inspection and manufacturing are provided below.

Inspection Groups

Group goal setting tended to be more effective than individual goal setting. Furthermore, the group procedure was preferred by the supervisors because it required less time and improved communication among the group members. Reductions of approximately 75 percent in both data transmittal and paperwork errors were obtained by the group procedure.

Before group goal setting, the average paperwork accuracy was 10 errors per unit. The group goal of five errors per unit was achieved five weeks after it was established and performance remained at approximately three errors per unit for the final month of the study.

In contrast, the inspectors who established a more difficult goal by individual goal setting failed to achieve their objective. The general trend for the period of the study showed no improvement as a result of individual goal setting.

The group procedure was initially more effective than individual goal setting in reducing data transmittal errors; however, data for the last month of the study failed to show a significant difference between the two methods. The inspectors using the group method reached their goal during the first week following goal setting and maintained an error rate of less than 1 percent for one month; during the last four weeks of the study, the group's error rate increased sharply as the result of errors by two individuals. In contrast, the average individual goal of a 5 percent error rate was not obtained until the fifth week after goal setting, but an error rate of 3 to 4 percent was maintained during the final month of the study.

As a result of these findings and experience gained in training the supervisors the following tentative conclusions were drawn:

1. A clear-cut distinction between group and individual goal setting is not possible. Group members are generally reluctant to establish a group quality goal until each individual has information on his contribution to the group error rate.
2. Group goal setting is at least as effective as individual goal setting in terms of reduced error rates.
3. Supervisors generally prefer the group procedure since it requires less time and improves communication among the group members.
4. It is generally not feasible to establish group goals without devoting several sessions to the factors which limit group effectiveness; that is, some group problem solving should precede attempts at goal setting.

Initial Manufacturing Groups

The purpose of the second study was to evaluate the combined effectiveness of group problem-solving and goal-setting processes in typical manufacturing operations. Five assistant supervisors in electronics assembly operations completed the training program and conducted meetings with their assembly groups.

Reliable data on quality and production performance were obtained for three of the five groups. Two months after initial goal setting, a cable assembly group had increased its output from 12 to 17 cables per week and reduced defect rates from 0.8 to less than 0.3 defect per cable. A change from a stationized line to individual responsibility for a complete unit in one circuit board assembly group reduced production time from 5.5 to 4.2 hours per board and reduced defect rates by 50 percent. Goal setting and problem solving in a second circuit board assembly line doubled the group's output of acceptable boards. This resulted in a cost savings of more than $25,000 during the six-month period following the problem solving/goal setting process.

The success of these initial studies led to a continuing growth in the application and development of this approach.

Supervisory Skill and Attitude

In recent applications the effectiveness of participative management was found to be highly dependent on the supervisor's attitude towards his people and his skill in conducting group sessions. Simply holding meetings did not necessarily result in high levels of participation or the desired performance improvements. To improve motivation the supervisor had to increase the extent to which people were actually involved in decisions that affected their jobs. For increased motivation to result in better performance the supervisor also had to provide required methods and job instructions. The importance of supervisory skill and attitude is illustrated below by the findings from four supervisor-employee groups.

In two electronic module assembly groups output and level of participation were observed for two months or more. Both groups performed essentially the same type of work and both supervisors had received training in the use of group techniques. However, independent evaluations by two psychologists indicated that one supervisor obtained a high level of group participation, whereas the other supervisor obtained very little, if any, participation. The supervisor of the low participation group tended to be defensive in meetings and discouraged participation by his lack of interest in the problems and solutions raised by his people. The results for these two groups are shown in Figure 13.6. The high participation group showed a significant increase in the number of good modules produced per unit of time, whereas the low participation group showed a steady decline in this measure. The supervisor of the low participation group terminated his data collection effort at the end of the fourth two-week period.

The second illustration involves the performance of one group under two different supervisors. This group inspected and touched up photoresistive material before the etching process for etched circuit boards. The performance of this group during an 18-week period is shown in Figure 13.7. During the second week the photoresistive material used in the touch-up operation was replaced with a new material found subsequently to be of a poorer grade. Adequate touch-up required several applications of this material and, consequently, increased the amount of time required for this operation. By the fourth week supervisor A had received his initial training in group techniques and had initiated a program of group goal setting and problem solving. From the fifth through the eighth week, a significant increase in performance was obtained.

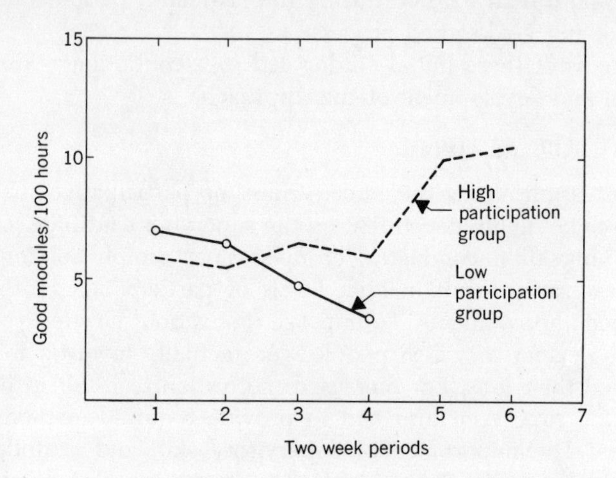

Figure 13.6 Quality performance of high and low participation module assembly groups.

During the eighth week supervisor A was replaced by supervisor B who also had completed the eight-week training program in group techniques. Although supervisor B conducted group sessions with his people, these sessions did not result in a high level of participation. Supervisor B used the meetings to stress the importance of high production rates rather than to involve his people in the solution of group problems. Production rates immediately decreased and stayed at a reduced level. This low level of output persisted under supervisor B from the ninth week through the eighteenth week, even though an improved photoresist material was provided for the touch-up operators during the eleventh week. The group meetings were continued, however, and the supervisor received individual coaching from a psychologist before and after each meeting. This process finally did produce changes in the supervisor's attitude and behavior. These changes resulted in an increased output of 5.2 circuits per hour during the twentieth week. This was equal to the level achieved by supervisor A.

Goal-Oriented Problem Solving

Previous research has shown that motivation is increased when employees set their own goals. Results for two groups of computer assemblers indicated, however, that employee involvement in the establishment of production goals is not critical. In these two groups the performance objectives were established by the supervisor at the level required to meet

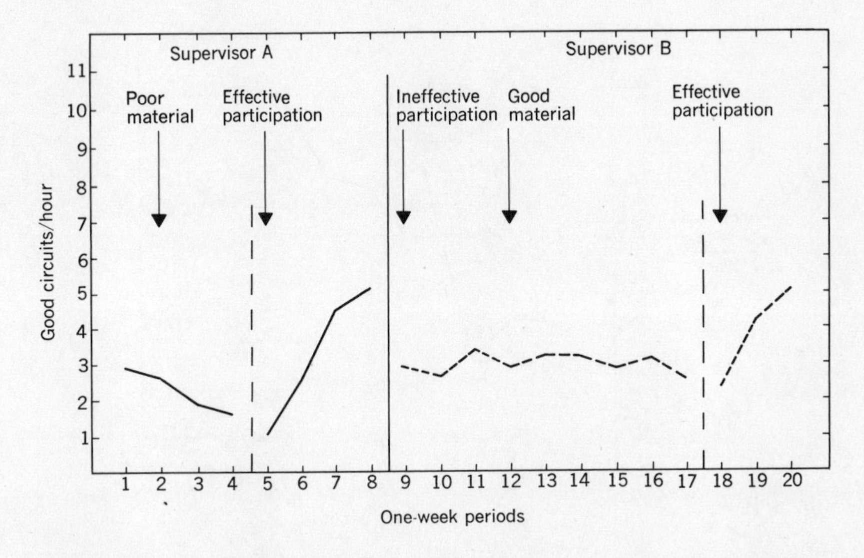

Figure 13.7 Impact of supervisory style on photo touch-up performance.

schedule commitments. In his initial group meetings the supervisor presented data on past performance, stated his new quality and output objectives, and explained why it was important to meet these goals. He then asked for a discussion of the problems that they would have to solve in order to meet the new commitments. Thus attention was directed immediately toward goal-oriented problem-solving activities instead of spending several meetings in the joint determination of group objectives.

The performance of the Read Head Assembly group had been relatively stable at about 300 acceptable parts per week. With no addition of personnel, the group output increased to 620 acceptable units per week during the first month of group problem solving meetings and exceeded the goal of 600 units per week during the remaining six-month evaluation period. Output exceeded 750 units per week during two of the last six months. Similar but less spectacular improvements were made by the Write Head Assembly group. The impact of this approach is illustrated in Figures 13.8 and 13.9. These relatively dramatic increases in quality of performance appear to result from two major factors. The first was the increased motivation resulting from greater participation in matters affecting the computer assembly job. The second was the actual job improvements that came out of the group problem-solving sessions. A number of specific problems impeding quality and quantity of production were identified and eliminated through the group's efforts.

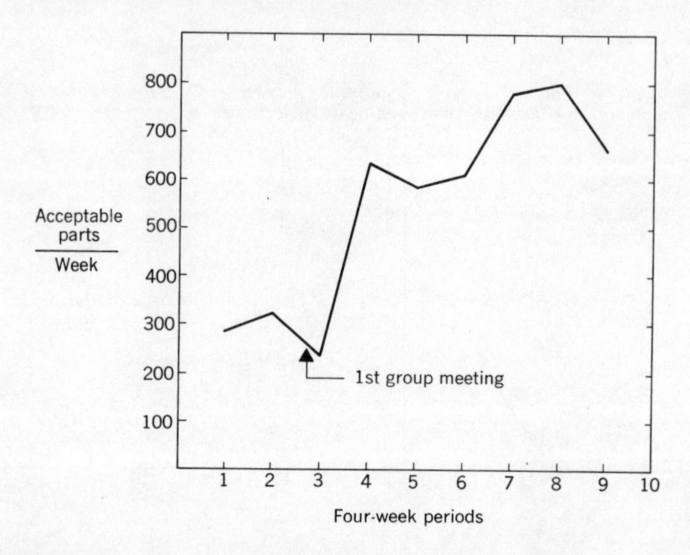

Figure 13.8 Improvement in Read Head quality and production as a result of goal-oriented problem solving.

Employee Attitudes

The relationship between employee attitudes and performance improvement was measured under four levels of group participation. Independent observations by two psychologists were used to classify the six groups in terms of the level of participation obtained by the supervisors. Two groups were assigned to each of the following categories: high, medium, and low participation. A fourth group representing the "no group participation" condition consisted of a random sample of electronics assembly personnel with no experience in group problem solving or goal setting. Group production output for a minimum of six weeks before and six weeks after group participation was analyzed to determine the percentage of change in performance for each of the four levels of participation. The percentage of positive attitudes toward the job was also computed for each of the four conditions.

Level of group participation was found to be positively related to both job attitudes and the amount of performance improvement. The score for job attitudes was twice as favorable for the high participation condition as it was for the individuals who had not participated in group sessions. With regard to the observed change in performance, no significant change was noted for the low participation group. Average performance improvements of 45 and 90 percent were obtained for the medium and

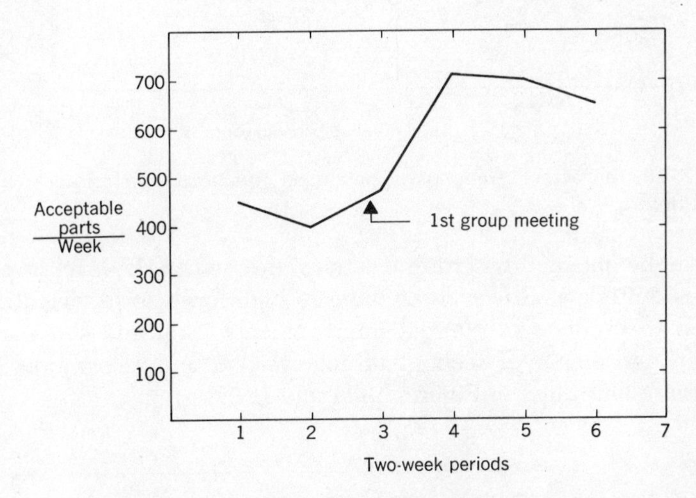

Figure 13.9 Improvement in Write Head quality and production as a result of goal-oriented problem solving.

high participation groups, respectively. These findings strongly support the idea that effective group participation can produce the changes in employee attitudes toward their jobs that are required for improved performance. The results of this analysis are shown in Figure 13.10.

The results were also analyzed to determine the impact of group performance on six specific types of job attitude: (a) production goals, (b) problem solutions, (c) instructions and methods, (d) equipment and tools, (e) achievement, and (f) recognition. The pattern of job attitudes towards problem solutions, achievement, and recognition were very similar across all levels of group participation. In each case there was no improvement with low participation and substantial improvements for medium and high participation. In general, the findings indicate that significant improvement in attitudes toward equipment and job instruc-

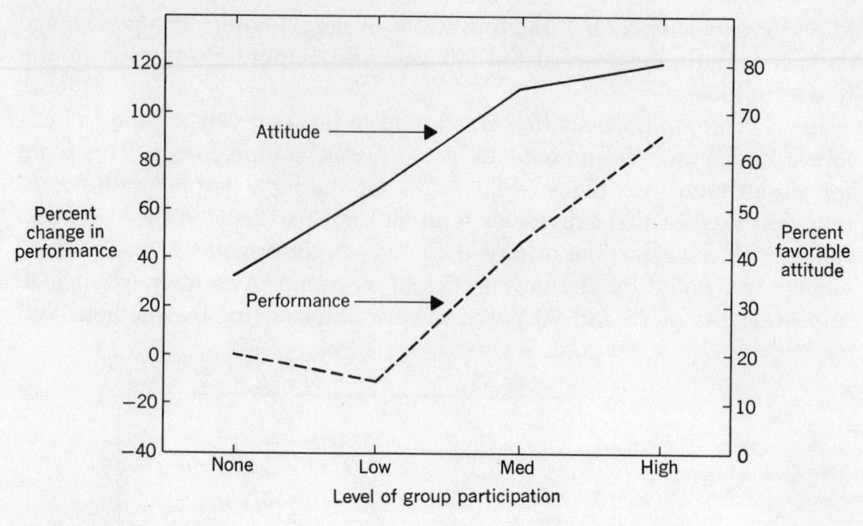

Figure 13.10 Impact of group participation on production performance and employee attitude.

tions may be obtained by group meetings, even when the level of participation is quite low. However, medium to high levels of participation are required to produce substantial improvements in attitudes toward job problems and employee feelings of achievement and recognition. These findings are illustrated in Figures 13.11 and 13.12.

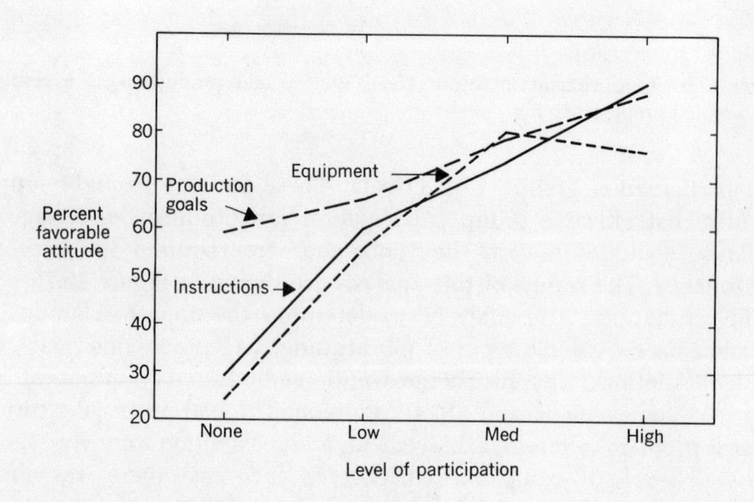

Figure 13.11 Impact of group participation on employee attitudes toward instructions, equipment, and production goals.

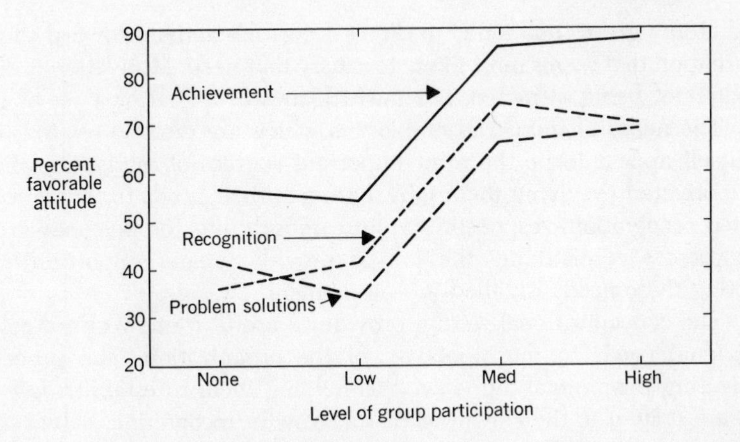

Figure 13.12 Impact of group participation on employee attitude about problem solutions, achievement, and recognition.

Human Factors Research and Engineering

The emphasis on motivation through job involvement and the positive results discussed in this chapter may lead one to conclude that motivation is the most powerful avenue toward improving quality performance. Therefore some perspective should be interjected at this point. These results and those reported elsewhere indicate that changes in quality performance resulting from improved motivation usually do not exceed an order of magnitude or two. Even improvements of this size include certain process changes initiated as a result of problem-solving activities. Changes in methods, tools, aids, information flow, and so on, have frequently resulted in improvements of several orders of magnitude [20]. Thus it should be recognized that although significant improvements can be made through motivational approaches, the greatest potential for improving quality performance in both manufacturing and quality assurance systems is in basic improvements in the systems themselves [21]. There are many examples of the identification of problems, the development of hypotheses concerning alternative designs, the experimental evaluation of design alternatives, and the implementation of the experimental results that have led to significant improvements in human performance.

SUMMARY

1. Motivation may be defined as the internal process that causes people to work toward goals that they feel will satisfy their needs. The strongest

need at any given time tends to direct a person's activity toward the goal or situation that seems most likely to satisfy that need. Motivation is clearly a matter of being attracted, not pushed, toward a goal or a set of goals.

2. The needs of industrial employees which are directly related to the job itself appear to be the most important sources of motivation. People are motivated by giving them jobs that meet their needs for achievement, earned recognition, responsibility, and opportunity for personal growth. The process of changing the job to provide greater opportunities for meeting these needs is called job enrichment.

3. The concept of goal setting provides a useful framework for relating individual needs to the objectives of the organization. The process of motivating people may be viewed as helping them to establish *job* goals that are related to their own needs for growth, recognition, achievement, and responsibility. One requirement for implementing this approach is a company-wide goal-setting system that establishes a meaningful hierarchy of job objectives for each level of management. This type of system helps the individual supervisor to establish personal subgoals with meaningful responsibility for each employee.

4. A high level of employee motivation is one of the rewards that a supervisor receives for doing his job effectively. One of the most critical aspects of the supervisor's job is the process by which he assigns work and evaluates results. High levels of motivation can generally be obtained through periodic meetings between the worker and his supervisor for the mutual determination of job goals, detailed planning of the tasks to be performed, and a joint review of progress. Cooperation between the supervisor and his employees in solving job problems is also essential.

5. One of the most effective ways to increase motivation is to allow employees to play an active role in decisions that affect their job. Unfortunately, participative techniques are often associated with permissive or weak management. To the contrary, research has shown that participative management creates stronger supervisors because they must develop clearly defined goals for their organization and valid ways of measuring progress towards these goals. Participative supervisors are also stronger in that they have a more detailed understanding of their operation, its problems, and the potential of their employees.

6. To increase and sustain high levels of motivation, it is essential to provide knowledge of performance results in a regular and timely manner. Research has resulted in the following recommendations for effective feedback in production operations: (a) when the defective part is identified, it should be returned as soon as possible to the responsible manufacturing supervisor, (b) the supervisor should discuss the error with the operator who produced it and make specific recommendations for avoiding

such defects in the future, (c) each operator should be allowed to perform his own rework, and (d) to provide a means of controlling feedback, random samples of rework items should be examined each week to determine what percentage of defective items are actually being returned to the responsible operator.

7. Recognition for achievement is particularly effective when it is provided in terms of more interesting assignments, increased responsibility, deserved promotions, and merit pay increases. To ensure maximum effectiveness reward practices should be consistent with the types of rewards generally desired by employees. Modern attitude measurement techniques provide a systematic means of developing effective reward systems.

8. The value of money as a motivator depends largely on what money represents to a person; money is not a universal motivator. In general, salary increases based on merit or other forms of pay that symbolize real achievement have a high degree of motivational value, but general pay increases, group insurance plans, and many other standard economic benefits which apply equally to all individuals in a company have little, if any, long-term motivational value.

9. To translate employee motivation into improved performance, supervisors should actively involve their people in developing new methods, tools, and procedures. This requires a joint consideration of the engineering and psychological factors that affect human performance. The concept is to establish *sociotechnical* systems that will enable employees and technology to produce the best results.

10. Supervisory development is a critical step in implementing a new sociotechnical system. The objective of this process is to obtain the changes in supervisory behavior and attitudes required to obtain a high level of employee support for new methods and requirements. A supervisory development program was described that included management development seminars, supervisory orientation sessions, employee group applications, individual coaching for each supervisor, and management evaluation sessions.

11. Applications of this supervisory development program have resulted in significant improvements in product quality and production output. The effectiveness of participative management is highly dependent, however, on the supervisor's attitude towards his people and skill in conducting group sessions. Simply holding meetings does not necessarily result in high levels of participation or the desired performance improvements. There appears to be a strong linear relationship between the level of group participation actually obtained and both a percentage of improvement and employee attitudes toward their jobs.

12. The emphasis on improved motivation through the use of participa-

tive techniques and the positive results reported may lead one to conclude that motivation is one of the most powerful avenues for improving product quality. While motivation is an essential ingredient for sustained improvement, the greatest potential for both manufacturing and quality assurance lies in the improvement of basic methods and techniques. By involving employees in the development of better ways of doing their jobs, we may obtain high levels of motivation and technical improvements with a high level of built-in employee acceptance.

REFERENCES

[1] Frederick Herzberg et al., *The Motivation to Work*. New York: Wiley, 1965.

[2] Frederick Herzberg, One More Time: How Do You Motivate Employees? *Harvard Business Review,* January–February 1968, 53–62.

[3] Carl A. Lindsay, Edmond Marks, and Leon Gorlow, The Herzberg Theory: A Critique and Reformulation. *J. appl. Psychol.,* 1967, **51,** 330–339.

[4] Charles L. Hughes, *Goal Setting*. New York: American Management Association, 1965.

[5] Charles L. Hulin and Milton R. Blood, Job Enlargement, Individual Differences, and Worker Responses. *Psychol. Bull.,* 1968, **69,** 41–55.

[6] Frederick Herzberg, One More Time: How Do You Motivate Employees? *loc. cit.*

[7] D. McGregor, An Uneasy Look at Performance Appraisal. *Harvard Business Review,* May–June 1957, 89–94.

[8] Edgar F. Huse, Putting in a Management Development Program That Works. *California Management Review,* Winter 1966, **9,** 73–80.

[9] E. F. Huse and E. Kay, Improving Employee Productivity Through Work Planning. In J. W. Blood (Ed.), *The Personnel Job in a Changing World.* New York: American Management Association, 1964, 298–315.

[10] Charles L. Hughes, Why Goal Oriented Performance Reviews Succeed and Fail. *Personnel J.,* June 1966.

[11] A. Marrow, Industrial Psychology Pays in This Plant. *Mod. Ind.,* 1948, **16,** 67.

[12] Frederick B. Chaney and Douglas H. Harris, *Human Factors Techniques for Quality Improvement.* A paper presented to the 20th Annual ASQC Technical Conference, 1–3 June 1966.

[13] Lyman W. Porter and Edward E. Lawler, III, What Job Attitudes Tell About Motivation. *Harvard Business Review,* January–February 1968, 118–126.

[14] J. M. Juran, Quality Problems, Remedies and Nostrums. *Ind. Qual. Control,* 1966, **22,** 647–653.

[15] M. S. Myers and E. D. Weed, *Behavior Change—A Case Study.* A paper presented to the 1965 meeting of the American Psychological Association.

[16] R. J. Pierce and S. C. Streep, Successful Motivation Programs. *Ind. Qual. Control,* June 1966, 654–658.

[17] R. C. Albrook, Participative Management: Time for a Second Look. *Fortune,* May, 166–200.

[18] Donald L. Wass, Teams of Texans Learn to Save Millions. *Training in Business and Industry,* November 1967.

[19] F. B. Chaney, Personnel Training and Evaluation for Quality Motivation. *Quality Motivation Handbook*. Milwaukee: ASQC, 1967, 91–106.
[20] R. C. Rook, *Motivation and Human Error*, Report SC-TM-65-135. Sandia Corporation, Albuquerque, New Mexico, 1965.
[21] J. M. Juran, Operator Errors—Time For A New Look. *Quality Progress*, 1968, **1**, 9–11.

INDEX